DESIGN EMPATHY AND CONTEXTUAL AWARENESS

Wayne K. Li

Frames of Reference for
the 21st-Century Creative

LAURENCE KING PUBLISHING

PREVIOUS PAGE: A beautiful mural? Colorful artistic texture?

LEFT: Taken out of context, the image on the previous page might be perceived as an arresting piece of street art. But seen in full, is it evidence for arrest? At what point do the lines between artistic expression and vandalism blur?

First published in Great Britain in 2025 by

Laurence King
An imprint of Quercus Editions Ltd
Carmelite House
50 Victoria Embankment
London EC4Y 0DZ

An Hachette UK company
The authorized representative in the EEA is Hachette Ireland,
8 Castlecourt Centre, Castlenock Road, Castlenock, Dublin 15,
D15 YF6A, Ireland (email: info@hbgi.ie)

Copyright © 2025 Wayne K. Li

The moral right of Wayne K. Li to be identified as the author of this work has been asserted in accordance with the Copyright, Designs and Patents Act, 1988.

All rights reserved. No part of this publication may be reproduced or transmitted in any form or by any means, electronic or mechanical, including photocopy, recording, or any information storage and retrieval system, without permission in writing from the publisher.

A CIP catalogue record for this book is available from the British Library

TPB ISBN 978-1-52943-821-5
EBOOK ISBN 978-1-52943-822-2

Quercus Editions Ltd hereby exclude all liability to the extent permitted by law for any errors or omissions in this book and for any loss, damage or expense (whether direct or indirect) suffered by a third party relying on any information contained in this book.

10 9 8 7 6 5 4 3 2 1

Design: Courtney Garvin
Project editor: Angela Koo
Commissioning editor: Liz Faber

Printed and bound in Dubai

Papers used by Quercus are from well-managed forests and other responsible sources.

Contents

Introduction		4
The Purpose of This Book		12

1.0 LAYING A FOUNDATION: DESIGN AS A WHOLE-BRAIN ACTIVITY — 14

- 1.1 An Interview with Scott Taylor — 16
- 1.2 What We Learn and How We Think — 24
- 1.3 Frames of Mind and Frames of Meaning — 32
- 1.4 Design Process — 46
  - Tool: Stakeholder Matrix — 70
  - Tool: Participation Strategy — 74
- 1.5 Case Studies — 76

2.0 DESIGN BEHAVIORS: DESIGN AS A WHOLE-LIFE ACTIVITY — 80

- 2.1 Behavioral Approaches — 82
- 2.2 Design Empathy — 84
- 2.3 Contextual Awareness — 86
  - Tool: AEIOU — 88
- 2.4 Creativity and Craft — 90
  - Tool: The Elements of Art — 92
- 2.5 Rapid Iteration — 98
- 2.6 Entrepreneurial Sustainability — 102
  - Tool: Hierarchy and Principles of Design — 106

3.0 EMPATHY: COGNITIVE AND EMOTIONAL EMPATHY, AND COMPASSIONATE CONCERN — 112

- 3.1 An Interview with David Kelley — 114
- 3.2 Needs versus Solutions: Reframing around People — 122
  - Tool: Empathy Map — 126
  - Tool: Maslow's Hierarchy of Needs — 128
- 3.3 Cognitive Empathy and Positionality — 130
  - Tool: Mind Maps and Social-Identity Maps — 134
- 3.4 Emotional Empathy — 138
  - Tool: Method Acting for Design — 140
- 3.5 Compassionate Concern — 144
  - Tool: Analogous Research — 146
- 3.6 Narratives and Interviewing — 148
  - Tool: The Open-Ended, Semi-Structured Interview — 152
  - Tool: Journey Mapping — 156

4.0 CONTEXTUAL AWARENESS: CULTIVATING YOUR DESIGNER'S SIXTH SENSE — 158

- 4.1 An Interview with Christina Choi — 160
- 4.2 Expanding Your Frame of View — 168
  - Tool: Powers of 10 — 172
- 4.3 Trends, Technology, and Culture — 176
- 4.4 Organizing Human Needs — 180
  - Tool: Common to Qualifier — 182
  - Tool: Affinity Mapping and 2x2 Matrices — 184
- 4.5 Timelines and Mapping — 186
  - Tool: 9 Windows — 190
  - Tool: Era Analysis—Material Culture — 192
- 4.6 Design and Business Strategy — 196

From the Author	199
Cultivating a Holistic Approach	200
Bibliography	202
Index	204
Picture Credits	207
Acknowledgments	208

Introduction

ABOVE: An advertisement for Baker Electrics from *Life* magazine, 1913.

Did you know that electric vehicles existed before Tesla? In fact, electric cars have taken on many forms in the past. Along with steam-powered vehicles, these alternatively powered cars were in the majority in America in the very early 20th century, when the automobile was still in its infancy. The honor for the first widely mass-produced electric vehicle goes to the Baker Motor Vehicle Company, founded in 1899 in Cleveland, Ohio. Starting with two models in 1904, Baker developed and expanded its offerings to include trucks and commercial vehicles, offering up to 17 models by 1910.

The Baker Electric vehicle of 1910 had several advantages, offering a quiet, comfortable ride (with no noise or vibrations coming from the electric motor, in contrast to the combustion engines of the time), a roomy interior (the electric motor was also smaller), and controls that were easier to operate. The early gas-powered cars of this era all had hand cranks to turn the engine over; these were very hard to turn and often caused injuries.

ABOVE: A 1915 advertisement for Rauch & Lang and Baker Electrics, promoting electric cars that were designed for both utility and pleasure.

However, the disadvantages of the electric car—including reduced power, a more limited range, and a higher cost—meant that it was primarily suited to (and marketed at) well-to-do women. At the time, vehicles were still seen as luxury items, generally within reach of only the wealthy. They lacked power steering and power brakes, so using the controls took some effort. Because of these issues, the first electric vehicles were seen as "ladies" cars. Similar to how we might view a golf cart today, easier-to-operate electric vehicles were to be used by wealthy women to make short trips to visit friends and tour their estates. In short, the Baker Electric was marketed by leaning into the weaknesses of the electric technology.

It was the mass-production techniques pioneered by Henry Ford and the invention of the electric starter motor by Charles Kettering that doomed the Baker to obsolescence. Where the Baker Electric cost around $2,800 (around $89,000 today), the Ford Model T was only $850 ($27,000) and it would continue to go down in price with improvements in mass production, reaching $260 by the 1920s. Once the electric starter motor was put into combustion engine vehicles, along with the developing gasoline infrastructure (gasoline in jugs sold at blacksmith shops and garages were popularized during this period), these developments made the mainstream adoption of the gas vehicle inevitable.

Another attempt at reviving the electric car came about as the result of the OPEC (Organization of Petroleum Exporting Countries) oil crisis in the US in the late 1970s. Responding to the shortage and high price of gasoline, Congress passed legislation to encourage the development of electric vehicles. Along with the development of both the Toyota Prius (a gasoline-electric hybrid powertrain) and Honda's electric EV Plus in 1997, General Motor's EV1 heralded a new era of electric vehicle availability. I was designing at IDEO around this time, and they were looking into the design of the inductive, paddle-charging station that accompanied this vehicle.

The EV1 was ahead of its time in many ways. Its regenerative braking technology and aerodynamic bodywork helped eke out every last ounce of electrical energy in its 16.5 kWh battery. As a technology demonstration of a commuter vehicle with an all-electric powertrain, it heralded a new era of electrically powered transportation.

However, the product still failed in the general market. Despite celebrity endorsements at the time from Mel Gibson, Ed Begley Jr., and Tom Hanks, major characteristics of this vehicle—like an average 80-mile (130-km) range, enough seating for only two occupants, a suggested cost of $33,995 ($66,700 today), and a driving dynamic that was sufficient but not exemplary (0–60 of 8.4 sec)—there wasn't enough there to convince customers to switch from their gas-powered vehicles. Additionally, due to a lack of infrastructure, there was no reliable national network of chargers to help recharge the vehicle along the way or when reaching a destination, so it was always doomed to fail.

Fast forward to the fall of 1998, when I had graduated from university with my engineering and fine arts (design) degrees, and had begun working as a vehicle-integration engineer at Ford Motor Company. They were very quick to show new employees their electric-technology vehicle: a 1996 Ford Ranger outfitted to run on electric power.

As an onboarding function for new employees, we were all invited to drive several production and development vehicles around the short handling course in Dearborn, MI. From a Mustang and an Aston Martin DB7 to the electric Ranger, we were able to experience these different vehicles firsthand. For the Ranger, the electric powertrain demonstrated a strong amount of power and quick acceleration, one of the hallmarks of an electric powertrain, which has almost instant-on torque characteristics. However, the ironic part of this technology demonstration was that the use case of the pickup, to carry items in its truck bed, was invalidated by filling the entire truck bed with around 12 lead acid batteries. There was no longer a reason to drive an electric truck because the cargo area was already completely occupied. I was impressed at the conversion but also puzzled about the odd engineering choice and made a mental note of it.

I had the sense, perhaps wrongly, that it was as if both GM and Ford—even with societal pressure and incentives from the government—weren't really interested in creating an electric vehicle that would succeed. They were both leaning into the disadvantages of an electric powertrain, either accentuating a very small range, or showcasing that lead-acid batteries (which

LEFT: GM's Electric EV1 c.1997.

LEFT: The 1996 Ford Ranger. This vehicle was used as a testbed for electric technologies.

ABOVE: The Tesla Roadster (2008–12). Martin Eberhardt's startup grafted an electric powertrain using lithium-ion batteries to a Lotus Elise underbody, creating an extremely light and fast electric sports car.

both vehicles employed) were heavy and bulky. They weren't trying to meet the demands of the market. Instead, they tried to build a technology demonstration to prove that electric vehicles were not a viable business model.

Why would this be? It's just my conjecture, but perhaps they didn't want to invest financially in the research and development of this new technology or in human resources, hiring new engineers specializing in this area. It could also have been due to the fact that car manufacturers don't sell cars directly to the public; they have partners through car dealerships who might have been greatly upset if the revenue-generating model of oil changes and tune-ups (spark plug changes and radiator coolant flushes) were made obsolete all of a sudden. (Electric vehicles don't have the same regularly scheduled maintenance and so are easier to maintain.) It could also be that they had partners in the oil and gas industry who would have had an incentive to keep selling gasoline for the products that used them.

In the late summer of 2004, I was a graduate student in product design and also a vehicle interaction designer for Volkswagen's Electronic Research Lab in Silicon Valley. Earlier that year, I held a teaching assistantship to Bill Moggridge, one of the fathers of computer-interaction design and the designer of the first laptop computer (the Compass) for Grid Systems in 1981. Through this relationship, Bill knew that I had experience in designing and developing cars. One day, he invited me to meet a neighbor of his, Martin Eberhard, who was developing his own electric vehicle in the form of a small startup company.

I met Martin, along with Bill, and we chatted about several topics around this new electric vehicle in his "garage." He asked me about interior automotive design and lighting technologies used in current vehicles. We also talked about which parts would provide a distinctive style over the costs of implementing them—particularly headlights, tail lights, and interior components, considering his bodywork was modified from a Lotus Elise. I described my observations on the current state of electric cars and the fact that they didn't seem to cater to the market very well.

Martin's views of the situation confirmed some of my own thoughts. The established car makers really didn't have any incentive to invest large sums of money in changing their business model, especially if it meant alienating their dealership partners, who would control their ability to sell vehicles and make a profit. Their business, built over a century, was not interested in change, and had become conservative. Their existing technology was proven and consistent, and their sheer size made them risk-averse. Thus their "attempts" at electrification were more exercises to demonstrate to Congress that electric cars weren't yet viable than an earnest effort to innovate and serve customers in a new way.

Martin described the ethos behind the development of his car, which contrasted greatly with the motives of larger, more conservative automakers. Firstly, instead of designing into the disadvantages of an electric powertrain on purpose, lean into the positives of the technology (instant torque and rechargeable; given alternative energy sources, a fossil fuel-free transport). Given the right engineering, there was no need to make excuses that an electric vehicle was inferior to gasoline. Finally, begin with an audience in mind. In the heart of the Bay Area, Palo Alto and the surrounding suburbs had many affluent car enthusiasts who owned Porsches, Priuses, and Ferraris. There was a need to demonstrate a virtuous lifestyle of environmental conscientiousness but coupled with a more selfish need to own a status symbol that was the pinnacle of performance. How could you cater to those tech-company executives who felt guilty driving their sports cars, polluting the planet (at 12–15 mpg [19–25 kpg])? Could you design a carbon-neutral option that could beat a Porsche off the line?

Being inspired by and catering to a specific human audience with a real human need breathed character and purpose into the car. Martin's original Tesla Roadster succeeded by having a distinct point of view, which mirrored and resonated with the needs of its audience. By catering to this audience, any disadvantages, such as range, were minimized. This customer doesn't often drive a Porsche on cross-country road trips. Rather, it's used for short blasts to the local taco shack, around a mountain pass and back. The range would never be an issue because it was doubtful that they'd leave the suburbs. Instead of picking Delco and Panasonic lead-acid batteries like those used in the EV1 and Ranger, Martin chose the latest battery chemistry technology available—lithium-ion batteries (also from Panasonic). Lithium-ion had a battery density four to five times greater than lead acid, which meant you could either use a motor much more powerful than the EV1 or have a range five times greater. Martin—along with Lotus engineers who'd developed the chassis and underbody, and a handful of designers in Menlo Park and Los Angeles—chose a mix of the two. Instead of an 80-mile (130-km) range in the EV1, the Tesla tripled its range to 244 miles (392 km).

Instead of a 137 hp motor in the EV1, the Roadster had a 248 hp motor that gave it a 0–60 time under four seconds, twice as fast as the EV1. The vehicle's Lotus underpinnings were also lighter than the EV1.

The rest is history. Because of the Tesla Roadster's success, even selling a handful of units (around 500) generated enough revenue and awareness among its most ardent customers that they became evangelists for the company's product. They set in motion the cultural acceptance for wider adoption. When Elon Musk took over Tesla as chairman and Martin moved to the advisory board, plans to scale the company with the Model S, and then Model 3, signaled the ascent of the company from startup to mainstream public company. In October 2019, Tesla as a company surpassed the value of General Motors. While no one can predict the future performance of a company, at the time of writing, Tesla's market capitalization is approximately 12 times that of General Motor's.

Why did this happen? Why is it that in some cases some people fail to create products that sell, while others find an audience that gladly evangelizes their work and almost literally sells the product for them? In engineering education, students are often taught to analyze the primary function of a product. While this is a good idea in theory, it's also wholly insufficient and belies the complexity in making successful products. If a product fails at its primary function, it isn't sold to the public. All cars can transport you from one place to another. If they failed at this, they wouldn't be cars. However, it's the secondary functions, the ones that affect the character of the product, that determine the financial successes of Baker, GM, Ford, and Tesla.

There is a difference between an EV1 and a Tesla Roadster, Toyota Corolla, or a Ferrari. Differences like power, driving dynamic, styling, status, and societal perception are secondary functions of the vehicle that determine whether the product resonates with its audience. In this case, Tesla started with a human condition: resolving guilt from driving high-performance vehicles and "green" virtue signaling to others. From there, Martin chose the appropriate technology (lithium-ion) to cater to the performance metrics that resonated with the customer. Further, by staging his first offering to the public as a "halo" vehicle, a high-end luxury sports car capable of beating Porsches versus a commuter car, he helped establish a brand recognition of his vehicle being the pinnacle of electric performance, which would then trickle down to all the lower models. These lower models would then be made in greater quantities, matching the size of an increasing customer base, and generating the mainstream revenue required to grow the company in stages.

How might one be able to understand and develop a knack for making the right decisions to create products that are truly impactful, that help shape society? There are many ways to develop your mind and skills in this way. Five design behaviors can be cultivated through practice, reflection, and refinement that can greatly contribute to your product's success. By practicing certain mindsets and being in the right frame of mind at the right point in your creative process, you can more quickly and more consistently create impactful products that will help move society forward.

"BY PRACTICING CERTAIN MINDSETS AND BEING IN THE RIGHT FRAME OF MIND ... YOU CAN MORE QUICKLY AND MORE CONSISTENTLY CREATE IMPACTFUL PRODUCTS THAT WILL HELP MOVE SOCIETY FORWARD."

The Purpose of This Book

Hello reader. Thank you for picking up this reference book on product design. I hope that I can serve as an agreeable guide down this path of discovery, engagement, and creativity. I invite your presence and hope that after reading this text you'll feel it was a journey worth taking.

Thanks again. And now, let's begin …

WHO IS THIS BOOK FOR?
Whether you're making something in a workshop, drawing the next kitchen appliance, laying out a web page, writing that next screenplay, or mulling over a breakthrough idea that will revolutionize the world, this book will help you achieve your creative potential and make your designs more poignant, more equitable, more fun, and hopefully, more successful in the market.

Though this book is intended for a university student in product design, or industrial design, perhaps well into their second- or third-year studio course, it will also be a helpful resource for any creative professional (fashion, graphic, or environmental designers; musicians, playwrights, inventors, entrepreneurs), or those who consider themselves creatives in search of a way to develop their ideas.

WHAT IS ITS INTENTION?
This book introduces the topic of design behaviors, which, as I'll demonstrate throughout, can be practiced and thereby refined. These behaviors tend to use particular mindsets (both cognitive and emotional) to help move your work forward. The purpose of this book is to demystify the "fuzzy" front end of the design process, where you're in the earliest stages of figuring out your design and who you're creating it for. This is the place where formative design research methods intermingle with business trends and marketing data analytics. The book will cover two behaviors: design empathy and contextual awareness.

Design is neither a "trade" skill nor a personal "visionary" lightning-strike event, but rather a way of living. By practicing a way of working and engaging with the world, you can improve your ability to create social impact. This practice involves exploring, engaging, creating, and evaluating your ideas as a way to arrive at something meaningful, not only for yourself but for an audience too. By conscientiously applying mindsets and practices at certain phases in your work, you can drastically improve upon the results of your creative endeavors, and speed up their adoption and viability in the market.

A WORD ON TERMINOLOGY
It's worth quickly mentioning the use of the term "product" in this book. Having studied fine arts, engineering, and product and industrial design, I tend to have a very loose notion of what a product is. My definition would be something that's a "product of your imagination." An industrial designer might think of a product as a mass-produced good for consumption by the public, whereas for a software engineer it will be a piece of software. My definition encompasses both of these, but it's much broader, including anything that the imagination of a creative person can dream up. So whether it's a consumer hard good, soft good (textile), piece of software, visual art, public policy, screenplay, or interpretative dance experience, these are all "products" in my mind.

How do I use this book?

The book is divided into four chapters, with interviews and a practical feature called "tools" interspersed throughout, all carefully designed for ease of navigation. While this text was conceived as a reference, it's also intended to build a community of practitioners around design behaviors, so QR codes have also been included, allowing you to contribute your voice to our collective dialogue.

1. Color-coded chapters: Each chapter is coded with a colored strip on the left-hand side of specific pages, making it easy to distinguish where you are in the book, and to refer back to previous sections if needed.

2. Texture-coded tools: The "tools" use both color and texture for quick recognition. Each tool has a short introduction, instructions on how to use it, and a summary of its advantages (and disadvantages). You'll also find templates to download and use in your design practice.

3. Interviews: Included are interviews with three prominent professionals, sharing notes on design, case studies, and anecdotes that also reflect my thoughts and experiences in design, providing some real-world context to what you're learning.

Download   Upload

4. QR codes: Each tool includes two QR codes. The Download code can be used to grab templates or worksheets that will help you complete each task. Then I invite you to use the Upload code to post your practice of the tools, establishing a dialogue with others on the same journey.

THE PURPOSE OF THIS BOOK | 13

1
LAYING A FOUNDATION
Design as a Whole-Brain Activity

DESIGNING WITH INTENTION IS A HOLISTIC ENDEAVOR. WE MUST UNDERSTAND OURSELVES BEFORE WE CAN PAY ATTENTION TO OTHERS.

1.1 | An Interview with Scott Taylor
1.2 | What We Learn and How We Think
1.3 | Frames of Mind and Frames of Meaning
1.4 | Design Process
    Tool: Stakeholder Matrix
    Tool: Participation Strategy
1.5 | Case Studies

1.1 | An Interview with Scott Taylor

Neuroscience and business

Scott Taylor is a professor of organizational behavior and the Arthur M. Blank Endowed Chair for Values-Based Leadership at Babson College. He's also a research fellow with the Coaching Research Lab at Case Western Reserve University, and a core member of the Consortium for Research on Emotional Intelligence in Organizations (CREIO).

The primary focus of his research is leader assessment and development. He enjoys studying the various methods organizations use to assess and develop their leaders, evaluating the effectiveness of these methods, and using his findings to develop new methods and technologies for improved outcomes.

In the course of carrying out my research into design behaviors, I was introduced to Scott Taylor, whose theories on developing leadership in business mirror my own theories on mental "frames" in design processes. Scott had observed a design exercise that I ran through executive education of senior leadership at a national home-improvement company and we started discussing what proved to be our similar views on, and approaches to, leading business teams.

In our conversation, Scott cited his colleague and long-time co-author Richard Boyatzis, who coined two states of being in business leadership—a positive emotional attractor (PEA) state and a negative emotional attractor (NEA) state. Each state consists of three interconnected neurophysiological aspects: the emotions, hormones, and cognition. Both states are important in leadership, but at the same time, they can not coexist under normal conditions; one cannot activate a positive and a negative emotional state at the same time. For someone to be a successful leader, they must understand when to tailor their state of mind based on the person in front of them and the context of the situation.

Taylor: As to emotion—positive and negative—we've learned a lot about its importance over the last 25, 30 years. Barbara Fredrickson, for example, was one of the pioneers in looking at positivity. Martin Seligman kind of shifted psychology; that's where positive psychology came from. But we know that when someone is in a positive emotional state, what's going on with their body, their hormones, and their cognition is very different than when in a negative state.

And what you have last is the cognitive state, which we like to refer to as the empathetic network, versus the analytical network. We're not talking about left-brain, right-brain pop psychology—we're actually talking about networks in the brain. Tony Jack, my colleague, doesn't think it's very accurate, but the technical term is the "task-positive network", versus the default-mode network, and these three states—cognitive, emotional, and hormonal—impact one another, play off each other. So we call them PEA and NEA.

In regards to cognition, the brain has two states, deductive and inductive, which will be covered later in this chapter. The two reasoning approaches similarly cannot be called upon at the same time. Scott spoke about this further:

Taylor: What we know about the two cognitive states, which you just alluded to—and Tony was one of the first to show this, I think, way back in 2010—is that they're antagonistic. They don't fire at the same time.

Now, we can increase our efficiency in going back and forth. And I would suspect ... I mean, just watching you teach, your ability to move quickly between these is very evident. You can identify where someone is and if they are in a problem-solving, focused, analytical state, and yet you can hyper-change over to this kind of social–emotional, connecting-with-others, prosocial behavior—self-aware in the broader sense of you, your own emotions, how you think you're impacting others.

(cont.) And so we believe that leaders, effective coaches, and effective teachers can move very quickly between these two states. But let's be clear—they are almost like magnetic poles. We can easily get stuck. And when we're basking in one, it's hard because of the antagonism to move to the other.

You might see this with deeply analytical engineering students. One roommate says, "Hey, let's go out, talk about going on dates," and the other is, like, "Leave me alone." That's that antagonism. Or, you're out having a great time with friends and then all of a sudden you get this text saying, "You guys screwed up in your calculations. Get into the engineering office right now!" And it's like, "Why are they bothering me? It's the weekend." That's that antagonism right between the two.

And so when you talk about creativity and innovation, that doesn't happen with negative emotion, the sympathetic nervous system and analytical thinking firing. If I'm heavily negative—if I walk into a classroom and say, "Hey, by the end of the semester, half of you aren't going to be here. You know, this is going to be the toughest course you ever had"—I'm shutting down learning, openness, and creativity, and I'm looking to the left and right. Then my students are my competition as opposed to my learning and development partners.

So the way in which we teach as professors, the way in which we lead, we're shutting down a lot of this creative potential because we're not aware of these distinctions. And when you become aware of them, you become tremendously influential. Because imagine, you're noticing your team is getting bogged down. Ideas aren't coming in. Well, we may be in the task-positive network or the analytical network. We have this negative emotion. We're feeling the stress of the sympathetic nervous system. Let's do something to alter one of those three states because we know it's going to affect the others.

So the emotion really matters here. The nature of the relationship really matters. If I don't like you and you're part of my team and we're supposed to be designing a new product, it's going to get in the way of all of that. The negative emotion impairs the ability to move back and forth. And if I'm feeling stressed out, and we're talking about the stress, that is shutting the system down. We're converging, we're isolating. We're not open, we're not receptive. And so it's much more than just the cognition part of it.

The other thing I would say is that we can leverage it. Yes, you can move back and forth quickly. But the best way we know so far to help people learn to do that is by actually putting them in these states. So when I do an executive program, I will intentionally design different activities (and I think you do the same thing) that are about forcing not just the cognitive switch but the emotional switch. I'll use an exercise where I ask them to think about someone that they've worked with who they think is an outstanding leader and who has really impacted their life. I'm generating gratitude, a sense of fulfillment, accomplishment, appreciation, and positive emotion. And that's by design because then I'm going to give them a question that's going to have them be very analytical and I need

"CREATIVITY AND INNOVATION, THAT DOESN'T HAPPEN WITH NEGATIVE EMOTION, THE SYMPATHETIC NERVOUS SYSTEM AND ANALYTICAL THINKING FIRING ... SHUTTING DOWN LEARNING, OPENNESS, AND CREATIVITY."

Effective leadership

Positive emotions  Parasympathetic nervous system  Empathetic network

POSITIVE EMOTIONAL ATTRACTOR (PEA)

A PEA mental state involves a threefold process of positive emotions (such as joy, hope, amusement), the parasympathetic nervous system, and the empathetic (default) network. All three do not have to be activated at once, but they often are. A PEA state invites openess and social bonding, and a learning orientation.

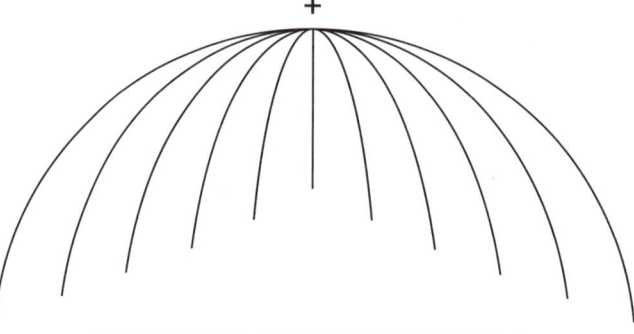
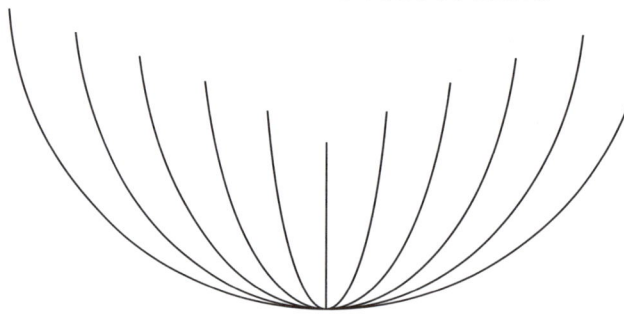

LEADERSHIP: TWO STATES OF BEING

Leadership begins with a personal vision, based on an ideal self or situation, which is adopted by others to become a shared vision. To create a personal or shared vision, you must be in a PEA state. While an NEA state is needed to move from vision to action, a person must spend significantly more time in PEA to achieve a sustained desired change.

Negative emotions  Sympathetic nervous system  Task-positive network

NEGATIVE EMOTIONAL ATTRACTOR (NEA)

The NEA state is characterized by negative emotions (like fear, anxiety, defensiveness), the task-positive network, and the sympathetic nervous system. An NEA state invites local attention, a prevention focus, and a performance orientation.

them to be open and receptive in that process; they'll be much more so if I can generate some gratitude and appreciation and positive emotion through that exercise.

Scott's research work sets a tone for this chapter, which itself will sound a refrain for all subsequent chapters: that designing with humanity in mind, being empathetic yourself, and understanding context will make you a better leader, entrepreneur, and designer. As a creative professional, your potential depends on your ability to avoid getting stuck in only "one way" of doing things, as comfortable as it may feel, and being able to challenge yourself to shift into new "frames" of thinking, whether those are positive or negative emotional states.

Finally, the sequence of these states of mind, and the speed with which you move between them, is critical in the development of your design work. Scott had a bit of advice regarding the staging of different states and how positive emotions can help confront negative ones.

> **Taylor:** The other thing is that the sequencing matters. You can't walk in and say, "Hey, we've got a problem we've got to solve." I mean, you can, but you're going to get much more mileage if you lead with a positive. Wayne, you're doing it through humor because of the style that you have and the way you do it; they're answering some tough questions and you're having them converge and solve and understand. But then, at the same time, you want them to be open and receptive. Well, humor speeds up that ability, right? If I'm laughing it shifts the emotional state, and then I've got to converge. And so you're enabling through the humor—laughing about the oranges and the juicer and what's going on: "Look at my colleague, he's kind of covered in orange juice …" All of this is an increase in that ability to move back and forth. I'll be, like, "My gosh, we're losing," but then I'll laugh because I've seen something funny. It's really quick.
>
> The other piece relating to that is the nature of the relationships being built through humor. I'm laughing with my colleagues, we're having a good time. We're doing something together that lets us know that a sense of relationship is really important to the other person.
>
> The nervous system, that positive hormonal system, creates resilience. It creates the desire to persist in moving forward through something. There are going to be times when it's going to be negative emotion. It's going to be stressful. We're going to have to converge and make tough decisions. But if we've got a relational camaraderie that we've built (and it doesn't take long to do that when we've had some positive experience to pull from) we're willing to go through that very differently—in a way that unites us, thinking together, and that increases creativity rather than separates and fragments it.

So sequence matters. As we look into design process, great leaders are able to quickly bring empathy and positive emotion to bear on the task at hand, before negative and analytical states are used. A creative professional should lead with empathy and creativity rather than with criticism and antagonism.

Taylor: One of the things in the Zoom world during COVID that we noticed in talking to leaders is that they said, "We're meeting more often because we can do it virtually." But what also happened is that people said, "Alright, we're not going to fatigue everybody here, so our meetings can be 60 minutes—get on, get up." They dropped all the laughing, all the connecting, checking in, and it was all task.

Well, that's what started to fragment the relational ability of the team. We advised them to start sharing some stories of what's going on because the walking down the hallway or hanging out before the meeting was gone. And you need that for PEA, otherwise, a meeting's going to be all NEA.

Look, you can be intentional. You can be aware of what's happening in a much more intentional way and then be intentional about what you're doing to shift those dynamics. And I think that's very true to the creative process that you're writing about. We can be more intentional and not just, "Well, whoever's got the kind of emotional social awareness to do that …" We can learn these capabilities.

I've interviewed thousands of leaders over the last 23 years, all over the world, and we've been to dozens of countries collectively. And I've found that outstanding leaders do two things. *One*, they are focused outward, on me—they create a connection to me in some way. It doesn't have to be intimate, but I feel connected to them. And *second*, they create an overall positive climate. It doesn't mean they don't swear or don't sometimes have a bad day. But overall, the environment they create is positive.

These two characteristics are incredibly important and we don't care what industry it is. I mean, what we're talking about is two characteristics that have nothing to do with all the factors of diversity out there. From Singapore to Springdale, Arkansas to Brazil, not for profit, for profit, high tech to healthcare, the answers are the same. And so when we start to analyze the data, these are what flesh out: these two characteristics, outward focused and positive emotion, whether directly connected to the PEA and NEA. And so the question is, how do you develop that in leaders?

"OUTSTANDING LEADERS DO TWO THINGS. *ONE*, THEY ARE FOCUSED OUTWARD, ON ME—THEY CREATE A CONNECTION TO ME IN SOME WAY ... AND *SECOND*, THEY CREATE AN OVERALL POSITIVE CLIMATE."

1.2 | What We Learn and How We Think

Our mindsets develop as we age and navigate the world. As your personality develops in childhood, your approach to work also develops.

How do you approach problems?

Are you methodical and deliberate? Do you seek out guidance before starting a project? Does that then affect what you think and how you deal with the world and larger problems? What we learn and how we think, and how comfortable we are in working through issues, is critical to creative endeavors. Creative blocks form when we're unwilling to change and like only to work on things with which we're familiar. Favoring tendencies and mindsets that are comfortable can lead to stagnation.

Alternatively, do you simply trust your gut? Do you try things out and see what happens and ask for permission later? Some people approach their work in a more improvisational manner, preferring to leave unknowns untouched, instead reacting and riffing off people while working collaboratively.

It turns out that many of the things we learn—our college majors—also favor particular approaches in their work. This is based on mindsets inherent in the disciplines themselves. So based on what we are familiar with, we may favor a particular approach. Over time, we may like to do things only a certain way. In the studio and in business, this can spell the death of your design or company.

> "In times of change, learners inherit the earth, while the learned find themselves beautifully equipped to deal with a world that no longer exists."
> American philosopher Eric Hoffer

Rational vs. intuitive approaches by discipline

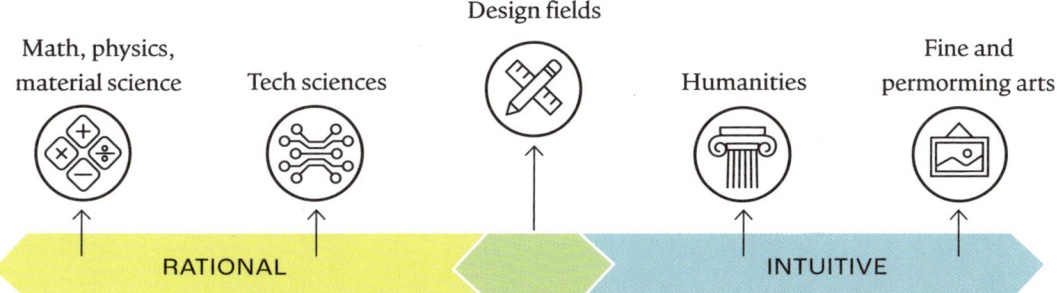

ABOVE: The disciplines that we take up tend to favor particular aspects of thinking and reasoning. Based on how comfortable we are with certain ways of thinking, we tend to gravitate to fields that suit our mindsets.

In the diagram above, you'll find that certain disciplines, such as the hard sciences (including math, physics, material science) favor more rational approaches, whereas the arts (fine arts, drama) are more intuitive in nature.

Why might this be? Why do some disciplines rely on a more rational mindset and others depend on more intuitive thinking? Is there something in the way that the subject is approached that tends to favor a specific methodology? Further, could someone who gravitates toward a specific body of knowledge—such as a scientist who might have been told in grade school that they were good at math and natural sciences—favor a particular way of doing things over others? Alternately, might someone who displayed musical talents early in life go on to approach problems with a more intuitive, instinctive mindset? What are the benefits and disadvantages of both approaches? And is it possible to combine the strengths of both?

You might have heard people described using the terms "right brain" and "left brain," where the right side of the brain tends to govern the left side of the body and vice versa. Further, many make an association between being "right-brained" and more artistic, and often left-handed, while people who are more logical or mathematically inclined tend to be thought of as right-handed and "left brained." Though it's not as simple as that characterization implies, it turns out that neuroscience does in fact provide a scientific basis for the idioms.

Researchers are currently exploring two types of reasoning involved in critical thinking: 1) inductive and 2) deductive reasoning. The reasoning approaches are also characterized as "fast" (Type I) and "slow" (Type II) thinking. Neuroscientists are looking at which areas of the brain are used for each type of reasoning and at how the two different processes contribute to decision-making. Let's look at each of them.

"BASED ON WHAT WE ARE FAMILIAR WITH, WE MAY FAVOR A PARTICULAR APPROACH. OVER TIME, WE MAY LIKE TO DO THINGS ONLY A CERTAIN WAY. IN THE STUDIO AND IN BUSINESS, THIS CAN SPELL THE DEATH OF YOUR DESIGN OR COMPANY."

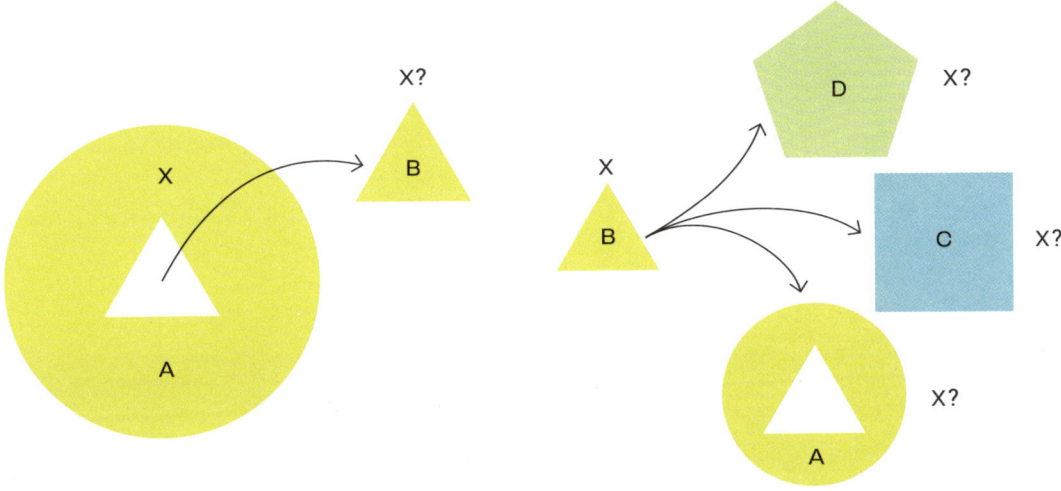

Deductive reasoning

A simple example of deductive reasoning would be something like, "All dogs have four legs. A German Shepherd is a dog. Do German Shepherds therefore have four legs?" Deductive reasoning tends to use a structure based on logic and rules to determine a characteristic of a subset of a known entity. If A has characteristic X, and B is part of A, then B shares characteristic X. This is generally a rational approach, and math and science tend to work this way—when testing a hypothesis, the scientific method will involve a control and an experimental group and they will be compared against each other for the desired characteristic in a repeatable lab setting.

Inductive reasoning

An example of inductive reasoning would be something like, "All German Shepherds are predisposed to nearsightedness. German Shepherds are dogs. Are all dogs predisposed to nearsightedness? Further, are all humans predisposed to nearsightedness? What about fish?" In this case, inductive reasoning is asking you to infer a characteristic from a small example and then apply it to a larger population. In order for this case to be correct, one would need a large set of data to validate the assumption, gathered from real-world observation.

Though there's some debate, researchers have posited that these contrasting reasoning approaches use different areas of the brain. Deductive reasoning is associated with the frontoparietal network (FPN; shown in blue on the diagram opposite). You can see from the MRI brain scans below that certain areas of the brain are active during deductive reasoning, involving the precuneus, cerebellum, and dorsolateral prefrontal cortex (DLPFC). Inductive reasoning, on the other hand, uses the cingulo-opercular network (CON; shown in yellow opposite), and it triggers different areas of the brain, involving the thalamus, dorsal anterior cingulate cortex (dACC), and the lateral prefrontal cortex (LPFC). Inductive reasoning involves a recognition of patterns, often drawn from comparing several different entities.

Brain activated during deductive reasoning

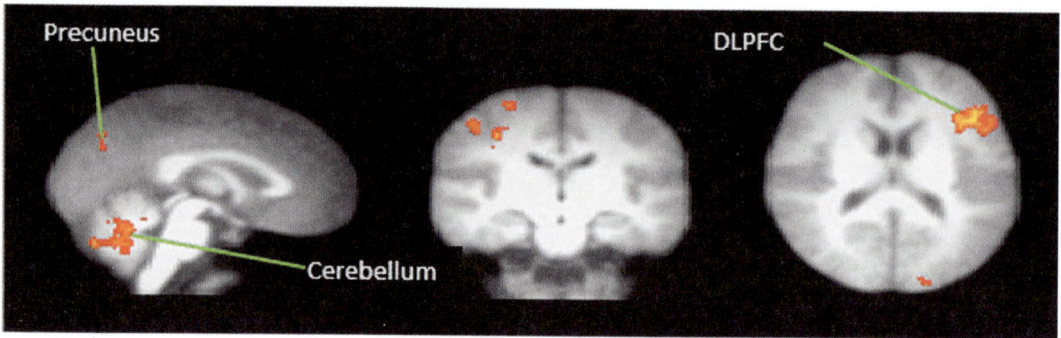

Brain activated during inductive reasoning

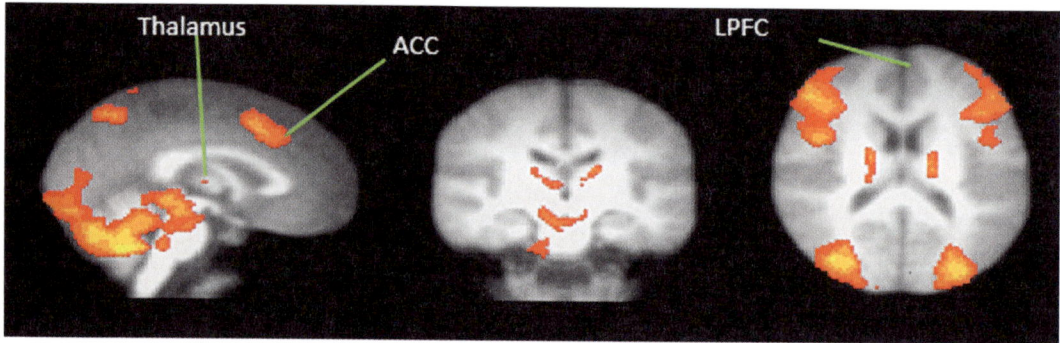

ABOVE: Different brain-activity patterns are associated with two different types of thinking: deductive (Type I fast) and inductive reasoning (Type II slow).

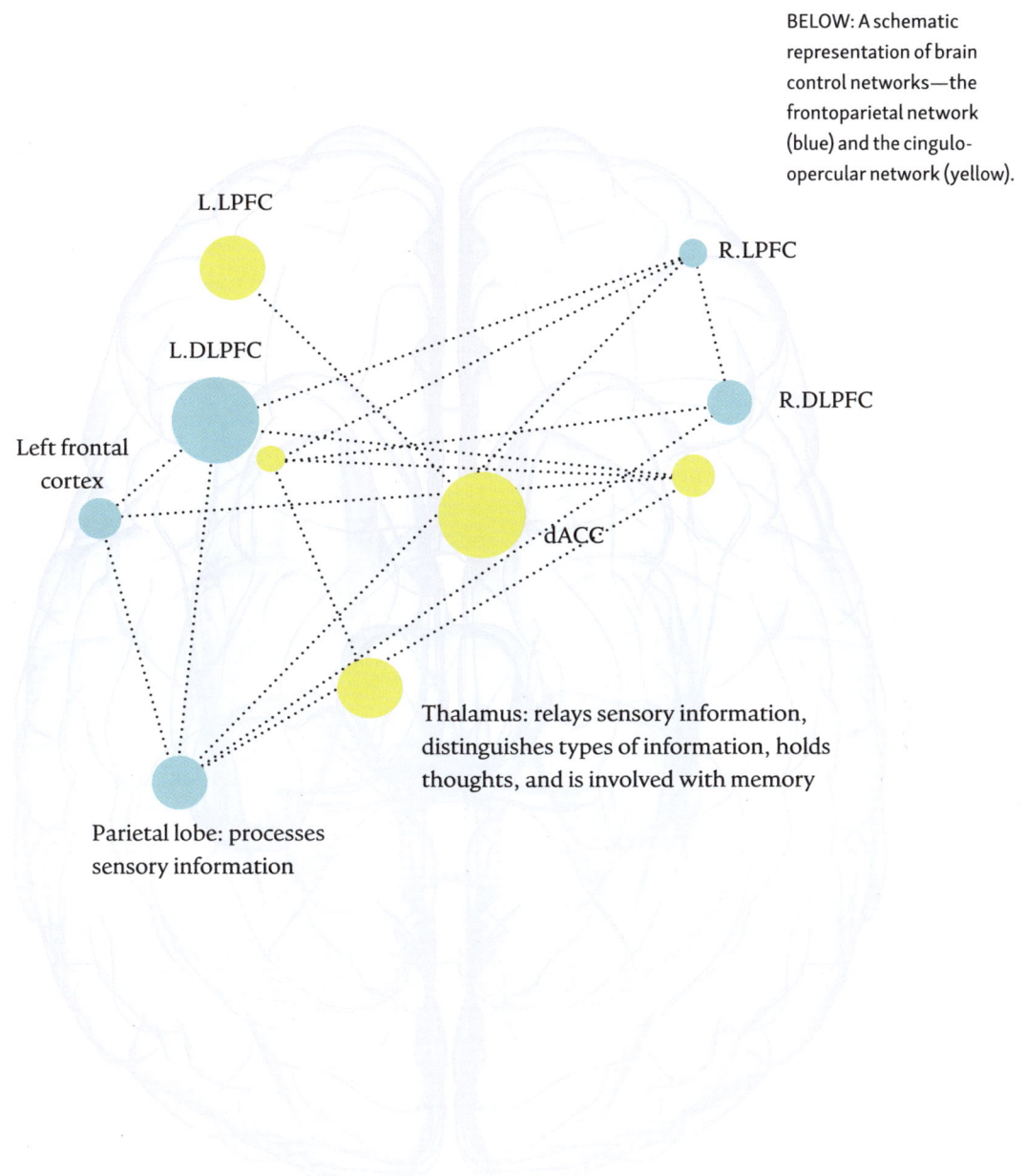

BELOW: A schematic representation of brain control networks—the frontoparietal network (blue) and the cingulo-opercular network (yellow).

Thalamus: relays sensory information, distinguishes types of information, holds thoughts, and is involved with memory

Parietal lobe: processes sensory information

Further, inductive reasoning tends to require more examples for reference to be gathered in the real world, so that a greater pattern of recognition can emerge. Also, note that the current prevailing theory (known as the "default-interventionist model") is that these two areas of the brain cannot be triggered simultaneously (which means that it's impossible to be both deductive and inductive at the same time).

Imagination, creativity, and intuitive approaches

We've covered a bit about which areas of the brain are used for different reasoning processes, so let's now expand upon that to see how it might relate to creativity.

In education, creativity can be taught—it is not a lightning-strike event or divine intervention. Sometimes, people claim that they're simply not creative, as if creativity is a natural gift that you're born with, or not. This is not true. As with critical thinking, creative thinking can be developed with practice, and it can be measured. It's entirely possible to improve upon your skills.

Research into creativity is being carried out in design, psychology, and neuroscience circles, and it's often hard to find a universally agreed-upon definition because it's often considered in relation to a particular field or industry of output (paintings by artists, screenplays by scriptwriters, algorithms by computer scientists, for example). For the purposes of this book, creativity can be thought of as the ability to produce ideas that are both novel (unique and original) and useful (appropriate to and resonant with a culture). It can also be thought of as a twofold process involving divergent thinking (also known as idea generation, or coming up with diverse concepts) and convergent thinking, or concept evaluation (reducing or grouping ideas through critique).

When we carry out a creative process, we're triggering several areas of the brain, and this involves both the reasoning approaches mentioned earlier, and the memory systems and experiences we've encountered to that point. For idea generation, we employ our imagination to visualize new possibilities, which often means letting our minds wander without direction or cognitive task (the default network). Our imagination helps us envision a future outcome, but in doing so, we also trigger the memory system into recalling the past—people, objects, and experiences—so as to construct a possible future relative to it. One region of the brain, the bilateral hippocampus, is active when memory, imagination, and creative thinking are engaged. This approach, of remembering and futuring, is known as constructive episodic simulation hypothesis (or simply, episodic memory), in which memory and imagination combine flexibly to remix people, places, and experiences. Idea generation and creation then tends to involve the default (at rest) network and the hippocampus.

Interestingly enough, our reasoning skills are also used for creative thinking. The DLPFC is associated with deductive reasoning, while the dACC is associated with inductive reasoning. Both are employed when we're in the executive network (high cognitive function) and occupied with creative activities like word association, divergent thinking, story generation, or musical improvisation. During such activities, these regions of the brain fire alternately to process information, evaluate the utility of novel ideas, and contemplate the relevance of those ideas. In this manner, we use both inductive and deductive reasoning to critique or evaluate concepts that spur further generation. The secret process for creativity, then, is the ability to switch quickly between episodic memory, futuring by analogy or comparison, and then an evaluative process using both inductive and deductive reasoning to spur the next round of idea generation.

An alternate approach that's worth mentioning here is the associative theory of creativity. This proposes that connections are forged between remotely associated concepts in semantic memory, which is a basis for "scientific" creativity. Semantic memory is the knowledge of words and concepts. What helps scientists become more creative is not only acquiring new conceptual knowledge (memory) but also modifying and reassociating the representation of the memory itself (reframing what the concept is associated with), which means a modification of the memory structure. The semantic memory networks of creative individuals have a "flexible" structure. Their networks have a greater number of connections between concepts, as well as shorter paths linking each concept.

When it comes to fostering creativity, a successful environment or culture must: encourage teamwork; invite freedom of thought, expression, and reflection without penalty; welcome idea generation without judgment; and enable ideas to be evaluated and judged, but not with the same breadth, or in the same session, as idea generation. Team members within a creative culture inherently trust one another, and they believe that ideas can come from anywhere, regardless of rank or status. Interactions are playful and improvisational, and build upon the diversity of the team's ideas.

You can often feel the vibe of a creative culture. It's lively and full of humor, and there's an earnest and open response as each new idea ventures forth. People are in the moment, in a collective flow state, and attentive to every other member of the group. They're not checked out, on the phone, or behind a laptop screen; they're checked in, playing off others' ideas and experiences in order to generate further ideas.

There's one final note to add here about the intuitive approach in relation to decision-making. There's a large group of neurons that fire outside the brain, known as the enteric nervous system (ENS) and located in the lining of your gastrointestinal system. The ENS can process information independently, and it communicates with the central nervous system and the anterior insula (AI) and the anterior cingulate cortex (ACC).

As mentioned before, the ACC is active in inductive reasoning, where we compare information to stored patterns based on previous experiences, often without conscious awareness. Based on the comparison, the ENS sends signals back to the brain, creating a feeling of right or wrong, which you experience as a gut feeling. "It just feels right" is a statement that an artist or poet might use to explain a creative decision. Your inductive reasoning is processing your gut feelings of right and wrong based on past experiences and comparing them against a current idea. This intuition is also often referred to as "trusting your gut instinct," or "going with your gut." Following this instinct is a more intuitive approach that points to a higher emotional intelligence quotient, developed by practicing empathetic mindsets and engaging with diverse people.

1.3 | Frames of Mind and Frames of Meaning

Let's begin with a quick question. Imagine you're in a classroom. The professor enters in a rush, with a look of urgency on their face. "I need your help solving this problem. We need to get an answer as soon as possible! We have to be correct, accurate, and get it right the first time. Mistakes will not be tolerated. Your grade depends on it!" The professor then procedes to writing the following on the blackboard: "4+4".

What is the first answer that comes to your mind, given the pressure of the situation?

The answer is eight, right? It's the fastest response and the most plausible answer.

In this case, due to the pressure to obtain the correct answer immediately, the human mind has a tendency to shift quickly into a deductive mindset. However, this is at odds with the slower inductive mind, whose role it is to try and grasp a deeper context.

But wait ... did I say this was a math problem?

Consider the situation for a moment. Why would a college professor need help answering a simple arithmetic problem? Further, what purpose would a professor have in asking a group of collegiate students a rudimentary elementary-school math problem? What would be the purpose of presenting a question like this to a highly educated audience? What do you think that professor would be gauging, based on the students' responses?

It turns out that many students, when under pressure, shift into a deductive mindset. They quickly fall into a pattern of behavior based on what they've encountered previously, making assumptions to simplify their thought processes so as to arrive at a possible "right" answer as quickly as possible. They fail to slow down and ask the inductive questions—why the question was asked, how they interpreted the language of what was asked, and the intention behind asking the question in the first place.

Let's look at the problem a little more closely. Look back at the sentence written on the board. In a deductive mindset, the mind often makes simplifying assumptions based on previous experience. This is very common in technical sciences, where engineering problems are often framed to fit within a category or type of situation, and simplifying assumptions are applied to them to speed up the process of arriving at an answer. Very common in geometry problems involving theorems and proofs, as well as engineering mechanics, simplifying assumptions are commonly phrased as, "Assume gravity is negligible," "Assume wind resistance doesn't exist," "Limit this case to a two-dimensional, planar situation," and so on. Of course, those simplifying assumptions are false. Gravity exists. Wind resistance exists on this planet. Objects tend to travel through three-dimensional space. The reason why those assumptions are made is to simplify the math needed to simulate the reality of the situation and make it easier to derive an answer that can be graded by a teaching assistant. Taking gravity, wind resistance, and three-dimensional cases into account would require a much higher level of mathematics than a first-year engineering student would have at their disposal. Unfortunately, because those more complicated edge cases are ignored, the way students learn tends to give them a false sense of confidence.

In this case, what were our simplifying assumptions?
1) Was the first symbol a numeral 4 or could it be interpreted as a different type of symbol?
2) Was the second symbol a "plus" sign, or could that also have a different interpretation?
3) Was this confined as a math problem or could we have interpreted the language in a different way?

Verbal

Let's rely on our more inductive mindset by imagining some alternate approaches to this visual language. What if this were not a math equation but more of a verbal sentence, such as C + A + T = CAT? Then the answer would be 44.

… or …

Visual

What if the symbols were to be used as a visual component of a drawing? In this case, the output of the sentence could be two ducks on a pond.

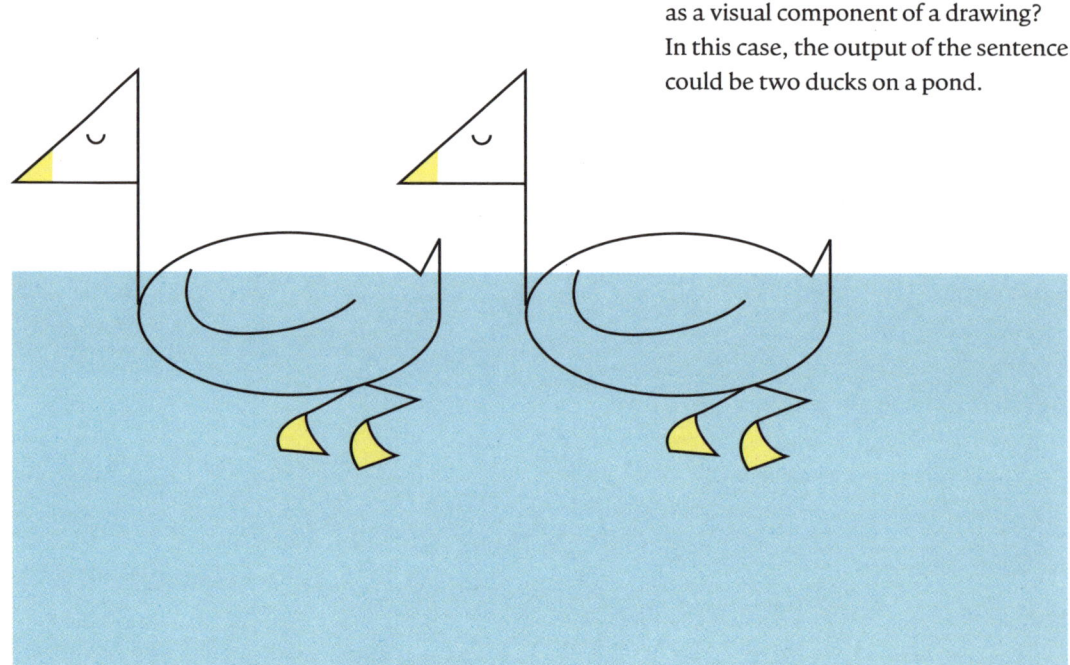

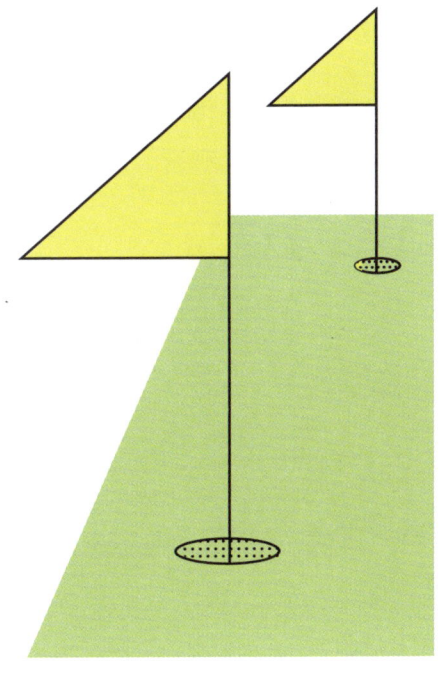

Perspective
What if the symbols were used to demonstrate the principle of perspective? Can we imagine the use of these symbols to indicate depth in a visual presentation? If this was the context of our problem, then the answer could be two holes of golf.

… or …

Combination
If this were a combination problem, such that only the symbol could be used to create something new, then the output of the equation could be a pinwheel.

"WHEN SOLVING COMPLEX PROBLEMS, HOW OFTEN DO WE 'JUMP TO SOLVE' USING OUR OWN EXPERTISE, VERSUS SEEKING AND EXPLORING OTHER POSSIBILITIES WITH EXPERTS, THE COMMUNITY, AND INDUSTRY FOR A MORE INCLUSIVE, RESONANT OUTCOME?"

Let's fall back to the implication of the original question. Why would a professor ask it? In this case, the responses from the students would allow him or her to gauge the class's creativity and inductive mindset capability. If a great number of students answered "eight," then the professor could conclude that the class was largely deductive. However, if a lot of the students examined the motive behind the question, or challenged the assumptions and asked clarifying questions, then their instructor could conclude that the class was more creative and aware of context.

It turns out that this is quite common in business. Business leaders and chief executive officers are often under immense pressure to present an immediate response to the problems that they're confronted with—and to make the correct decision in each case. The constant drumbeat from shareholders to maximize returns and provide maximum profits places a great emphasis on the need for instant results. Those leaders who default to knee-jerk reactions and deductive mindsets, however, will often mistake the type of problem they're confronted with. Further, without taking the time to contemplate the actual context of the problem and the myriad factors that created it, such leaders are then burdened with a solution that doesn't truly address the issue; as a result, they'll be confronted with the same problem in a subsequent quarter. When solving complex problems, how often do we "jump to solve" using our own expertise, versus seeking and exploring other possibilities with experts, the community, and industry for a more inclusive, resonant outcome?

We live in a social system. The products and services we create involve multiple systems and people. From factories and producers, to supply chains, distributors, and retail salespeople, the things we design will involve many different points of view. Most creators or technical experts mistake the type of problem they tend to face. Whether it's a simple system, a complex and organized system, or a complex and disorganized system, we creators need to understand that multiple mindsets—especially inductive and creative ones—are needed to frame the issues at hand. No one discipline carries the secret to solving the world's problems. Climate change, poverty, injustice, and inequality are complex, multimodal, and multifactor issues. The only way to address them will be for the world's designers, scientists, and other experts to come together and collaborate with an empathetic sense of common purpose.

> "Aim at Heaven and you will get Earth thrown in. Aim at Earth and you get neither."
> C. S. Lewis

Frames of mind

Unlocking your design potential requires understanding when a particular frame of mind can be more helpful to a situation. As mentioned in the previous section, if we rush to solve a problem due to the insecurity of not knowing the answer, our deductive mind steps in and often makes gross assumptions about the true nature of the problem. Or, when confronted with an unknown, do we slow down, and allow our intuitive and inductive minds to wander around previous experiences and tap into our unconscious experiences for inspiration or a different perspective on the issue?

We've already covered the neuroscientific basis for different ways of thinking—inductive versus deductive reasoning. We can also group ways of thinking into frames of mind, or mindsets, of which there are myriad categories. In design, we step into ways of thinking that are particularly adept at specific types of work. For our purposes, let's consider three mindsets and their use in design practice.

In the previous section, we looked at creativity and what's considered creative thinking. Consider inductive reasoning, an intuitive approach, and positive emotions as the creative mindset. The second would be the empathetic mindset, which requires perspective taking, emotional mirroring, compassion, and concern for others (see Chapter 3). The third is the critical mindset, which brings to bear logic, deductive reasoning, and analysis on the topic at hand.

Because these mindsets tend to use different areas of the brain that do not fire together, our ability to tackle complex problems and "design holistically" relies on our ability to shift fluidly between the three of them. In addition, our ability to identify the appropriate mindset for tackling a particular task is critical to a successful design outcome.

For example, when tasked to design something, you're confronted with a problem that exists out in the world. Let's imagine that you've been given the job of designing a new product for collecting rainwater in arid agricultural areas. You'd need to have an empathetic mindset to understand the issues that farmers of the region confront when dealing with water scarcity. Empathy would be required to understand the way they live, and the way they go about their business of farming the land.

When designing with the farmer and their input in mind, it would be best to operate with a creative mindset, in which positivity, intuition, and inductive reasoning—to stretch far and wide with creating ideas—would give you an advantage. You'd want to create a large number of ideas quickly, to be able to survey all the possibilities, then leverage them to arrive at the best outcome.

And finally, if you'd had several ideas built, and were testing different irrigation and water-collection systems in the field, you can see how a critical mindset would then come into play. Narrowing down those ideas in order to ascertain which were the most effective at a specific task (collecting water, or channeling water to the correct places, for example), or had the greatest impact on the lives of the farmers (highest crop yield per gallon of water), would be helpful for deciding which option to go ahead and produce.

As you start any design endeavor, be cognizant of the mindset that you're in and the way in which, during the project, you will switch between different mindsets. Are you stuck in a comfortable way of thinking? Do you avoid taking on a different mindset, or feel uncomfortable when you do? We've already mentioned that, based on what you've previously learned, you may not be comfortable thinking from a different point of view, or doing things that seem silly or weird. If you consider yourself introverted by nature, then talking and empathizing with a stranger may be difficult. Or, if you prefer logical and rational approaches, then being asked to be silly and create random ideas may feel foreign.

When you first start out, you may not be aware of the mindset that you're in. However, there are definitely signs that you'll pick up on as you practice design more. You'll know, for example, that you're being empathetic when you find you're truly listening to the person across from you, picking up on their emotions, and feeling genuinely curious about their life. In this state of mind you never make the conversation about yourself, and you're able to resist the temptation to interject in order to compare your situation to theirs. Often their emotional state will be contagious, so if you feel downtrodden when they are, or joyful when they're celebrating, then you know you're empathizing.

Creative mindsets feel playful, fun, and improvisational. You won't know what will come out next, or where anything is going, but you will feel present. There'll be a sense of trust among your teammates (if designing in a group), and you'll find yourselves hanging on each other's every word. It'll be a safe space for playful banter, and the thinking will run anywhere and be about anything. The mindspace will be supportive, allowing any idea to be voiced and heard. Even if the line of thinking moves off into unknown—even racy or offensive—territory, the space will welcome it, encourage it, and empower those with ideas that are counter to the mainstream.

Critical thinking, on the other hand, is a mindset that most people are already familiar with, and it's the mindset that the majority of people are comfortable using. It's also generally seen by academia as the most important approach to cultivate. Are you being logical, analyzing something, or are you employing comparative thinking (judging one idea against another on a common standard)? Are you trying to discern or judge a quality in something? Are you seeing if a behavior in a subset of data is consistent with a larger population? If so, you're in a critical mindset, where deduction and logical reasoning are at the forefront.

The important lesson to learn as a designer or creative professional is that you need to cultivate all frames of mind. Don't be content or comfortable in one mindset only; instead, learn to identify personal strengths and weaknesses in each, and encourage the use of all of them. Using your entire brain as you work will guarantee much better outcomes and a more engaging design experience.

Concrete versus abstract

Another concept we need to cover is the notion of concrete versus abstract ideas, which will be referenced throughout this book. Prevalent in art, this concept details the literalness of an idea. For example, a photorealistic still life or self-portrait can be considered concrete (a representational, literal translation), such as the Rembrandt self-portrait shown below. Though the brushwork, sense of light, and depth of field are masterful, the depiction is still a faithful rendition of the artist's visage.

On the other hand, if we take Duchamp's *Nude Descending a Staircase (No. 2)*, shown opposite, then the representation that we're looking at is an abstract (Cubist) painting of a human being.

Duchamp chose to abstract a particular quality about a human descending a staircase and represent that quality faithfully. In this painting, the quality of motion over time represents the limbs' motion as a sequence. Cubism as an art movement tends to have this quality, where

LEFT: Rembrandt van Rijn's self-portrait of 1659 was modeled on a famous portrait of the Italian diplomat Baldassare Castiglione, painted by Raphael in the previous century. Rembrandt used the pose, costume, and expression to present himself as a learned painter.

the cubes are detail images of a larger whole, meant to convey the experience of an object or person over time. In this way, Cubism heralded animation and the development of the first movie cameras in the early 20th century. Some people may find abstract art difficult to understand because they lack a frame of reference (knowledge of the quality chosen by the artist). However, once a viewer is able to grasp the painter's intention or thought process, then she gains a newfound appreciation for the work.

We can also take this concept beyond visual representation and apply it to human behaviors and the mindsets that drive them. For example, we can understand human behavior that's easily observable, and measurable in reality as concrete behavior. If a customer is checking out at a grocery store, then it's a simple matter to observe them putting items onto the conveyor belt ready to be swiped across a barcode scanner, then bagging and paying for them. These actions could be timed, recorded, and broken down into discrete steps.

However, there are also abstract behaviors at play that are not so easily observed. These may be actions that require some interpretation or piecing together of observable, more concrete activities. They may point to something that's not easily observable but is a relevant factor nevertheless. For example, if you notice that the person in the checkout line is nervously tapping their feet while waiting their turn, or that they pack their bags quickly relative to other customers, you might infer that their time is a limited resource. Or you might notice that the goods they're placing on the

ABOVE: Marcel Duchamp's 1912 *Nude Descending a Staircase (No. 2)* used abstraction to depict a sequence in time and movement. A controversial inclusion in the original Armory Show in New York in 1913, it is now held at the Philadelphia Museum of Art.

conveyor belt are all plain-packaged generic items rather than branded goods. This might lead you to believe that the shopper is a cost-conscious person who thinks branded goods are overpriced. Abstractions require the viewer to make an inference; they involve interpreting the situation and extracting a quality that seems authentic and/or important to the subject. There is inherently some judgment involved in this mindset. The main idea here is to understand what is faithful to the person and the situation. Abstract behaviors, such as making healthy food choices, shopping frugally, or rushing through activities, help point to a person's state of mind, which might not be so apparent on the surface.

One helpful way to distinguish between concrete and abstract behaviors is to think of them as a relationship: cause and symptom. Using the above example, being pressed for time is an abstract behavior (cause) that manifests itself in quick arm movements when packing items (symptom). This relationship is an interesting thought exercise because behaviors can continue to extend into further levels of abstraction. In this way, a cause can actually turn into a symptom of a much greater and/or more abstract cause. What causes someone to be in a hurried state of mind? Maybe their occupation dictates that they work long hours, leaving them very little personal time for things like grocery shopping. What caused their occupational choice (symptom)? Perhaps it was their need to maximize their earning potential (cause). You can see that earning potential or societal net worth is a very abstract concept, whereas performing a rapid arm movement is a relatively concrete one.

Ideas can also be concrete or abstract. They can be easily actionable and understood, such as a command like, "Please open this jar of pickles for me." It's easy to create a plan of action around a concrete idea. However, that plan of action will also have a very limited shelf life. Once the jar has been opened, the vacuum seal that caused its closure has been addressed, so the plan to open it the first time is only used once. The request for help to open a jar of pickles will not come again until this one is replaced with a new jar.

However, some ideas can be more abstract, such as, "I feel like I have no purpose today." Here, an action plan for finding a positive new direction in life would be an ongoing process. The longevity of this action plan would have to be assessed every day, to gauge whether or not it was providing the right guidance.

When dealing with abstraction, you're using a more intuitive mindset than a rational one. Your intuition as to what might have caused the symptom being displayed by another person will rely on your inductive reasoning ability and your learned experiences more than any simple, observable fact patterns in front of you. Also, because this requires both empathetic and creative mindsets, it's very possible to come up with multiple plausible theories on the abstracted cause. The only way to validate whether or not your intuition is right, is to engage the person and ask.

Frames of meaning

Imagine this childhood scenario: You come home after school to find a freshly baked plate of brownies in the kitchen. Yum! As you go to grab one and take a big bite, your mother walks through the doorway and says, "Don't eat those brownies! You'll ruin your dinner!" You walk off in a huff. A little later on, you return to the kitchen only to find your mom there eating those brownies herself. "Mom, I caught you red-handed. You're eating brownies before dinner!" Guiltily, she replies, "I'm your mother, don't do as I do … do as I say."

What causes this? Is your mom a hypocrite? Or, is there something deeper at play here? Everyone displays behaviors that may seem contradictory or hypocritical at times. This is entirely human. However, understanding the reasons in such cases requires deeper inspection. Your mother may have a particular intention when dictating your behavior, but a different set of rules may apply when dealing with her own. Why might she feel differently about her own behavior?

Let's imagine some of the factors that might contribute to this. Your mother may see you as a young person who hasn't yet learned how to manage their dietary choices, so she may be trying to teach you something. In relation to her own choices, though, she is already aware of any nutritional risks. Another point of view might be that because she's much older than you are, the risks of unhealthy foods aren't as severe; she may be comparing her mature years to your youth, where unhealthy eating still has the opportunity to compound over time. Yet another view might be that she sees herself in you, and is living vicariously through you, and thus would prefer you to live a better, healthier lifestyle than she has lived.

If we expound on that idea, why would parents want to live vicariously through their children? Why do they have children in the first place? Is living through your children a reason to have them? What are the joys of parenting? If we take the premise that one of the reasons parents have children is to live through their actions, then it makes sense that they'd not want their children to make the same mistakes they did—they'd want them to live better, more fruitful, longer lives. Seen from this perspective, your mother sees you not as a child but as an opportunity to live on. Selfishly or selflessly, she wants that life to extend as far into the future as possible. Not to be too morbid, but on her deathbed, who will she expect to be by her side? Who will she want to remember her at her funeral, or to talk about her with their own children, carrying on her memory? From this perspective, you're not her child, and the brownies are not an indulgent snack. The scenario can be reframed with an entirely different explanation—the brownies represent a legacy of cooking, and you represent her immortality.

The renowned psychologist Jerome Bruner coined a term, "intentional states," that helps illustrate this concept. Intentional states comprise pairs of "saying" and "doing" (action). They can be congruent (where what you say about what you do is a match) or incongruent (where saying what you do actually contradicts your actions). When we take several of these intentional states together, we begin to define our behavior in the world. For example, if we say we're going to work, and we do so. When we say we'll finish a task by Friday, and we do. If we say we'll deliver high-quality work in return for a just compensation, then we expect that compensation to be fair. Taken together, these intentional states can be considered the attitudes around work. This could be considered "acceptable work behavior," which would be arriving to work on time, delivering on promises, working on tasks diligently, and expecting equitable pay. Note, I would expand this definition to cover not only what we say, but also what we may be thinking or feeling. In this way, we have what could be called "say-do-think-feel" intentional states, and taken as a group, these states define an individual's behavior or a group's culture. Bruner would call this grouping a "frame of meaning." Our minds take several intentional states together to ascribe meaning to the world around us.

What is happening when what we say and what we do are at odds with each other? It turns out that there's a psychological basis for this—the ability to hold two different views of the world at the same time. The social-psychology term cognitive dissonance, or cognitive bias, is defined as a discomfort caused by holding conflicting ideas simultaneously. People instinctively want to reduce this dissonance, and they tend to do so by justifying, blaming, or denying. Cognitive dissonances mark changes in attitudes, beliefs, and actions. They are marked by having needs operating at different levels. When thinking of legacy and immortality, the long-term need to live a more perfect life will override a more selfish, short-term need to indulge in sweets.

When our "saying" contradicts our actions or is at odds with our thoughts or feelings, then unwritten rules and patterns begin to emerge, and these may comprise behaviors deemed unacceptable to a group of people. These are unacceptable meanings within culture and opportunities for design.

Design functions best when addressing contradictions within intentional states like these, when one group's frame of meaning is at odds with another's. The design opportunity lies in helping assuage these kinds of dissonances, to make life better, safer, more equitable, more fair, simpler, more interesting, or more fun, just to name a few.

You'll notice this in the public arena. When a politician or celebrity displays behavior that doesn't match the culture's intentional states, then they must rush out and make a public apology, explaining why they've not met our expectations. They can deny the behavior: "There's no evidence of me eating brownies!" They can try to reduce the impropriety: "Everyone eats brownies before dinner, it's perfectly normal!" Or they can turn to rationalization, pointing to an alternate option as mainstream behavior: "In my childhood, my family always ate sweet snacks before dinner. It was traditional to our family."

As you notice your own frames of meaning, and pay attention to the news or social-media feeds, note how yours may or may not match those of others. If we could provide a way to honor our parents while also enjoying a tasty snack together (without ruining dinner), then the world might be a happier place.

1.4 | Design Process

> "I love new clothes. If everyone could just wear new clothes everyday, I reckon depression wouldn't exist anymore."
> Sophie Kinsella, *Confessions of a Shopaholic*

Let's start with this simple sweatshirt. It's a fun fashion item that seems like it would be a quick buy in a trendy boutique.

One thing we can think of when we talk about empathy is resonance with an audience. When you look at this top, does it resonate with you? Is this something that's moved the needle? Is it a "must have" or is it merely something that tickles your fancy or piques your interest?

Now let's expand our view of it. What if I told you that this company's 300 designers put out 12,000 designs every year? What if I also mentioned that this company's model is to provide accessible high-end fashion to the masses? Do you think this is a good way to run a business?

It seems natural that a customer would want to be seen wearing clothes that the most affluent among us wear. But is this truly why we wear clothes? Do we get up in the morning wanting to look like a celebrity or social-media influencer? High-end fashion is expensive because it only caters to the top 0.5 percent of the population. Do the other 99.5 percent of us aspire to wear the exact same clothes as the richest, but in the form of copies made of much poorer-quality materials? This trend, known as "fast fashion," has been prevalent for the last two decades. Many companies have adopted it as a way of offering very quick turnarounds of large quantities of different designs for the sake of searching out ever smaller pockets of demand, targeting smaller and smaller groups of people. As creative professionals, we should think deeply about why we make things, who we make them for, and what their social impact will be.

1.4 DESIGN PROCESS | 47

"AS CREATIVE PROFESSIONALS, WE SHOULD THINK DEEPLY ABOUT WHY WE MAKE THINGS, WHO WE MAKE THEM FOR, AND WHAT THEIR SOCIAL IMPACT WILL BE."

What if I were also to tell you that around 90 percent of the clothes that the fashion industry produces end up in landfill? And that in this century these trends in clothing design push toward ever cheaper and ever quicker turnarounds, influencing customers to wear clothes for half as long as they did before, trashing them much faster than they did prior to the year 2000? What are the implications of these trends?

 90%

of the clothes that the fashion industry produces end up in landfill.

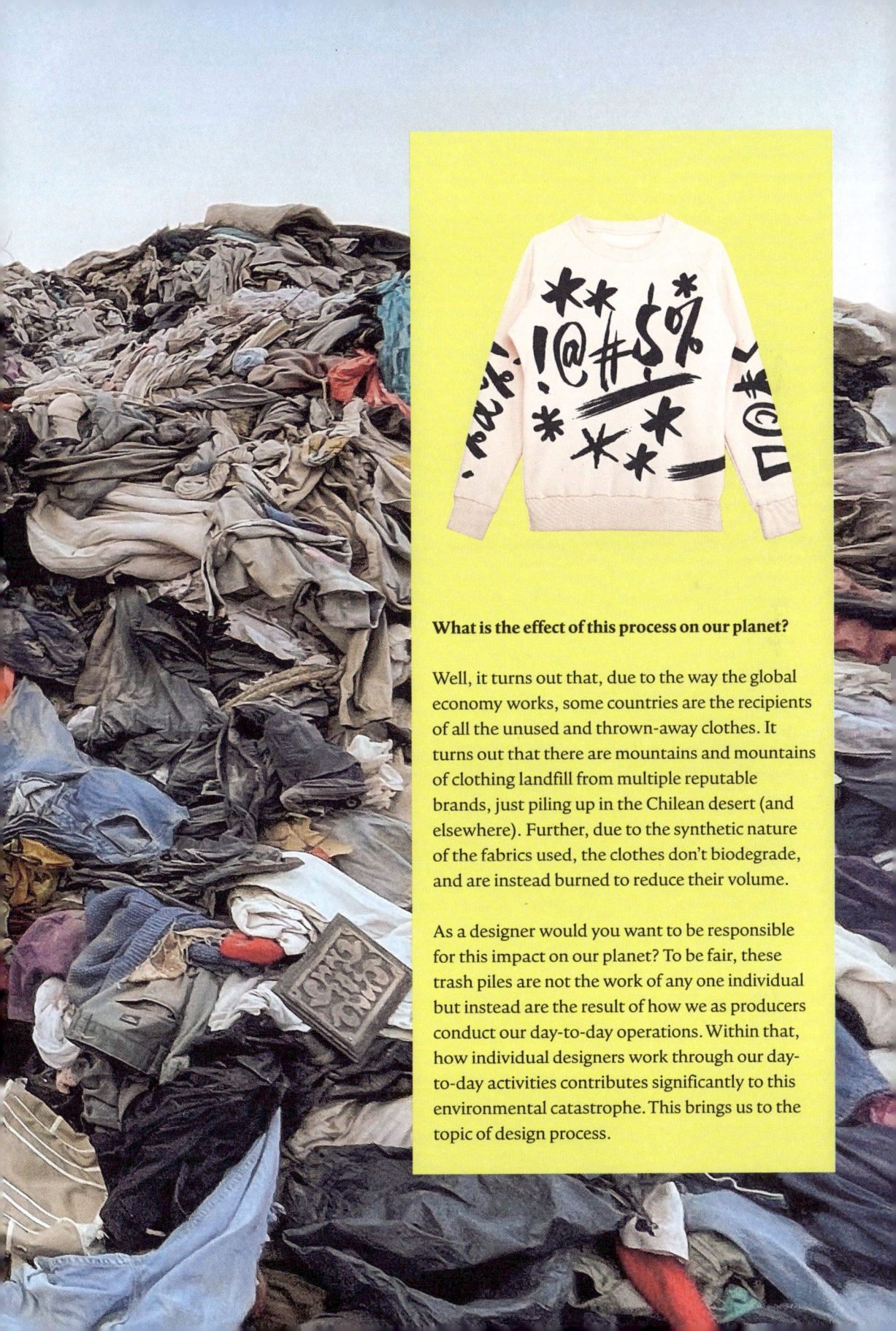

What is the effect of this process on our planet?

Well, it turns out that, due to the way the global economy works, some countries are the recipients of all the unused and thrown-away clothes. It turns out that there are mountains and mountains of clothing landfill from multiple reputable brands, just piling up in the Chilean desert (and elsewhere). Further, due to the synthetic nature of the fabrics used, the clothes don't biodegrade, and are instead burned to reduce their volume.

As a designer would you want to be responsible for this impact on our planet? To be fair, these trash piles are not the work of any one individual but instead are the result of how we as producers conduct our day-to-day operations. Within that, how individual designers work through our day-to-day activities contributes significantly to this environmental catastrophe. This brings us to the topic of design process.

A design-process example

Everyone is a little "d" designer, in that all of us will go through a design process. This is simply a set of phases and mindsets that a creative individual uses in order to understand a context and create some thing in the service of someone.

If you make eggs in the morning for your roommate then you are an "avian culinary designer." If you went to university to study the subject for four years, you'd earn a degree in it and you'd be a capital "D" designer (or a BSc ACD for short). However, what makes you a Michelin-starred chef is how well you learn, practice, and improve upon your design process.

Let's examine another industry in a fashion-adjacent realm. The product-development process that's featured over the next few pages comes from the retail home-decor industry. How might this organization's design process contribute to the fast-fashion trend?

The concept sketch below was created by a designer at a home furnishing and decor company—the kind of sketch that many designers within the company would draw to visualize what a room in a typical customer's home might look like, complete with the types of products that would be sold in a retail

environment to decorate that home. The image below displays some of that design intent translated into real production goods. From the picture frames, desk, hutch, and chair, to the clocks, enameled accessories, and matching bag and tape measure, the way in which these products would have been designed, produced, and sold would have been coordinated so as to evoke a particular emotion or tell a story to customers. The additional images shown here are examples of the diversity of products a retailer like this would offer.

BELOW: Products from a retail decor company, such as this task lamp, pocketwatch clock, and vintage phone, will coordinate with home-office furniture and accessories to tell a story or evoke an emotion.

The fashion industry and home-decor retail sectors operate on a seasonal timeline one year in advance. So in this case, if it were August, we would be designing the fall season for sale in the following year. Because home decor is a fashion industry, the seasons—spring, summer, fall, and winter—each have unique collections of products that are offered to the customer for a limited period. Some collections carry over from season to season as demand warrants. Other collections that are seasonal in nature tend to be offered only in that quarter, though highly successful seasonal products may be offered year over year. In all cases, demand from the customer and historical sales data determine when a product is removed from the shelves and no longer offered for sale, or if it continues to be in the stores year round.

The design process will then proceed in an established sequence of steps as follows:

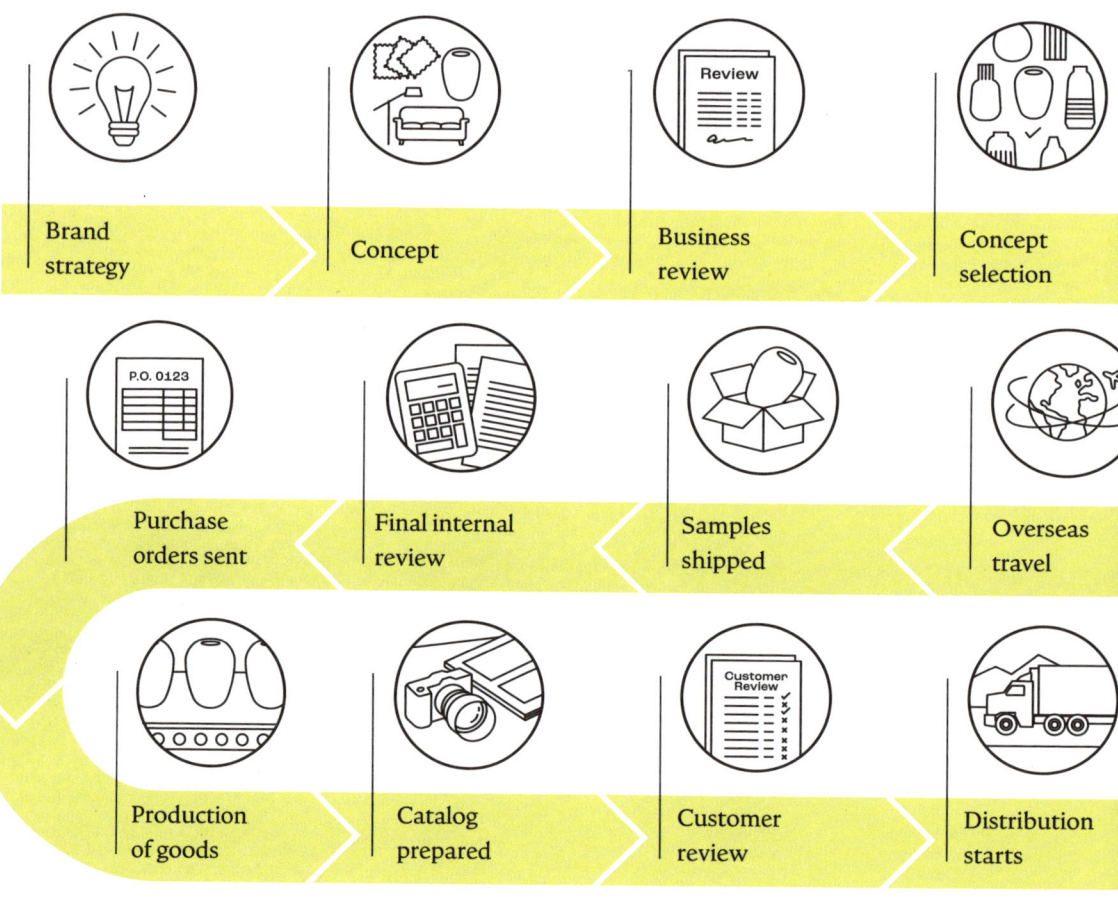

Brand strategy: *Day 1*
The process begins with the kickoff meeting, which examines the brand and is the first planning session for what the creative staff and executive management feel the stores should look like and what stories they should evoke.

Concept: *Day 21*
This leads into an initial concept meeting, where the creative staff pitch ideas, collect material samples, draw first sketches, and visualize what the individual pieces for the store and catalog might be. This is a time for the designers to give input as to the look of the season.

Business review: *Day 30*
In the next month or so it's the business office's turn. They will hold successive meetings, beginning with one dealing with, say, the sales of the current product in the stores, and then another based on the business requirements of the organization to offer a diverse assortment of

Final design & sourcing

First samples

Read and react to sales

product at the right price. Based on the sales history, for example, a merchant might say something like, "These five collections of vases sold poorly this season. In order to keep my assortment, I need five new vase collections." Note, the merchant is describing business requirements and not human needs. No one needs a vase. Rather, customers buy vases so that they can fulfill their needs, wants, and desires; they want to display flowers and arrange their living room in a way that pleases or impresses others.

This is an important distinction, and the phrasing is critical. Products or solutions—the things we make and the services that we provide—are all instantiations of our intent that we feel best address the human requirement. If we were to phrase the issue as, "Our customer needs new vases," then all we would create would be some type of turned pottery. However by phrasing the human requirement as the need to display flowers in a way that impresses others, the designer now opens up their creative space to things beyond a vase. What are all the different ways that we could display floral arrangements to impress others? A designer could create a mural composed of flower petals. Or they could create a wall of felt draped like a curtain and infused with succulent seeds; with a few waterings, that felt would turn into a living wall that could potentially generate more impact for the home at a retail price higher than a vase, while costing less to manufacture. There are countless solutions beyond vases alone.

The frame of meaning can limit our mind arbitrarily. Also note that the business needs are not human centered. The merchant doesn't need five new vase collections, she needs to generate more revenue than the lost sales of the five collections that underperformed. If one product could generate the missing revenue, then that would be all that she'd need to sell.

Concept selection: *Day 50*
At about two months in, the next meeting is a concept meeting that takes into account business requirements and the design narrative showcased as drawings and models in a retail and home environment (think of a production studio like a movie set, with a fake retail store and 12 areas mimicking the rooms of a home, with a kitchen, living room, bedroom, and so on). From here, the creative department and executive management walk the house and store and check them against the requirements that were established earlier. If a particular product or collection doesn't meet the standard sought by the merchant teams, then the designers are asked to generate more ideas and further refine their work—or to start over completely with a new concept for the five-vase collection requirement. Hopefully, the ideas reviewed by the team are acceptable and the design specification process can begin, with technical blueprints and computer-aided design drawings generated for manufacturing partners, aka "suppliers."

Final design and sourcing: *Day 71*
At this point, the design team will work with their sourcing strategy counterparts to identify which countries and manufacturing partners will be responsible for the production of the designs. The designs are then sent to each country where samples from the drawings are produced.

ABOVE: A product sample board (top) shows an assortment of botanicals, vases, candleholders, and Christmas ornaments, coordinated to tell the story of a season, while the product-development headquarters (bottom) provide a space dedicated to depicting both home and retail environments exactly as intended.

ABOVE: We live in a global economy, and shipping from the producing country to the destination port is part of the supply chain.

First samples: *Day 105*
Once the various suppliers receive the designs, they begin the "sampling" process. This is a collaboration between the supplier (which can be anything from an artisan shop all the way up to a mass-production factory) and the designer to execute on the design direction. The first samples, which are handcrafted by the supplier or made using soft tooling lines, are then air-freighted to the design department. The designer comments on the sample quality, execution, and aesthetic, then relays that via email/electronic platform back to the supplier. If any concessions or compromises are needed for production, the designer signs off on design-specification changes.

Overseas travel: *Day 130*
The next phase is one of overseas travel, during which the executive team and design managers travel to each supplier. This takes about three or four weeks and it may span from around 12 to 20 countries. The team's main aim is to assess the quality of the samples coming out from the suppliers and gauge whether or not they match the intent of the design drawings. At this point, if product is considered substandard—for either technical quality or aesthetic reasons—it is redesigned or adjusted. Or, in the worst-case scenario, the entire collection is scrapped and redesigned from scratch on the fly with the team in country.

"HOW COULD YOUR OWN PROCESS ... HELP RESHAPE HOW THE BUSINESS PRACTICES DESIGN?"

Samples shipped: *Day 169*
Once everything meets expectations, the first few samples of the suppliers' work are air-freighted back to headquarters to update the retail space and room settings.

Final internal review: *Day 180*
The last internal review takes place at about six months into the process. Now the retail store and room settings are populated with the products that have been made in perhaps 20 different countries, many using hand processes, and some using mass-production factories. This final review is the last checkpoint. This is when the same executive and creative teams, as well as all of the design, sourcing, merchant, and marketing teams who were not on the overseas travel, get their final say on the products for the season.

Purchase orders (P.O.s) sent: *Day 205*
From here, with production quotes from individual suppliers, financial allocations and purchase orders are generated. The retail manufacturer now arranges payment and authorizes the suppliers to begin mass production of the samples that have been reviewed and approved.

Production of goods: *Day 210*
The production of salable goods begins.

Catalog prepared: *Day 215*
At about the seventh month, the first runs of mass-production product are air-freighted to an actual store and to where the catalog is to be shot (usually a real residence). At the retail store, the company now holds its only external review with actual customers.

Customer review: *Day 220*
At about five months out from product launch, the company's most ardent customers are given an advance preview of product that will be in the stores later in the year. This is the only time that the home-decor industry gets customer feedback—at a point when all production monies have been allocated and there are few resources to change or improve product because the launch window has already started. What happens if customers don't respond positively to that advance preview? What chance does the business have to adapt or change those goods to meet the needs of the customer? Even though there have been several opportunities for the entire organization to review its work and make changes, if those assumptions of what truly resonates with the customer are wrong, there's no choice but to outlet (significantly reduce prices) or landfill those salable goods.

Distribution starts: *Day 245*
In the final stages of the process, the goods are loaded into the distribution channel and shipped by boat to the distribution point (intake port). Then they are transported to each retail location (by rail and road).

Goods for sale: *Day 365*
With 45 days to go before the product is offered to the public, the goods are shipped to each store. After a year of development, the goods are now available for public consumption: The floor of the stores are set and the business office gets ready to react to the sales of the new product (in addition to monitoring sales of existing items). The sales of this season are used to inform the process for the subsequent year and thus the process repeats and the cycle begins again. Also note that within the home-decor industry, each cycle takes place over a year, but the development for each season (spring, summer, fall, and winter, or holiday) is happening simultaneously. There are actually four cyles happening at the same time.

Conclusion

Now that you've seen how the fashion and home-decor industries carry out their design process, think critically about the steps they take. How do their actions contribute to their results? What's wrong with this process?

Do these businesses have models that are sustainable? Do they seek out human requirements in order to address needs within the market or do they try to force upon their customers a story or aesthetic that's of the moment, trendy, or out of touch? It is the design process used by fast-fashion companies that generates those mountains of unworn clothing piling up in South America and elsewhere. If you were an employee tasked to create the products

ABOVE: The retail environment, where products from multiple categories (furniture, textiles, storage, decor) are all presented in a coordinated design.

> "BY TRYING TO FORCE DEMAND ON CUSTOMERS ... INSTEAD OF ACTUALLY IDENTIFYING THE NEEDS OF THE POPULACE, FAST-FASHION [HAS] GENERATED LARGE AMOUNTS OF WASTE."

sold by a retailer, what would you do differently? How could your own process, used within a company, help reshape how the business practices design?

Further, because of the seasonal nature of the fashion and decor industries, retail businesses have separate product-development processes for each quarter running in tandem. A fashion designer will be running four simultaneous product-development processes at any point in a given year, which can be overwhelming and lead to burnout. It also encourages the corporate culture that birthed the fast-fashion trend in the first place. By trying to force demand on consumers by rapidly producing copies of haute couture instead of actually identifying the needs of the populace, fast-fashion companies have generated large amounts of waste. If only such businesses could recognize that their product-development processes are not conducive to understanding customers and creating products that delight and resonate with them, how might the business model change?

The product-design process: summarized and simplified

Let's take a look at these steps again in chronological order, with a variation on the previous flowchart. There are some phases (shown in yellow) in which the business is trying to understand the problem, investigating things such as sales history, customer demand, market trends, production capability, and changes in customer demographics. Other phases (in green) are creative ones, in which the team is generating ideas, visualizing solutions, generating blueprints and technical drawings, and fabricating samples and production-level goods. There are then other phases (blue) in which the outputs of the creative phases are assessed and analyzed against parameters important to the company. Finally, remember that this is a seasonal sector, so a designer in this organization would repeat this design process every 12 months (as indicated with the dotted lines). In the case of fashion, the designer would be juggling four simultaneous design processes—no easy task. Note that the steps here are not linear. In this diagram, creative phases are followed by analytical ones, and empathetic phases may kick off a development process but they also occur throughout development. This is to say that

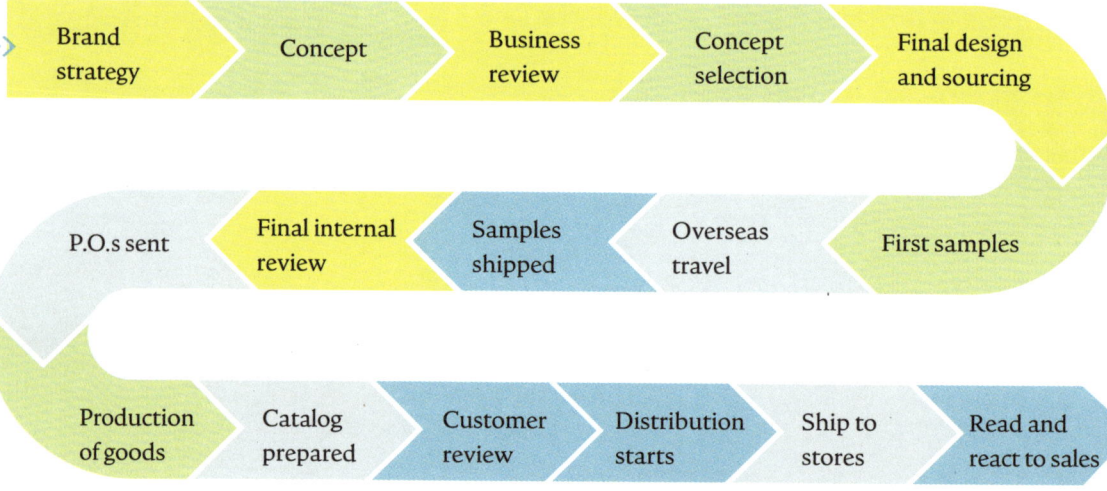

Phases of the design process

- Steps for understanding the problem
- Steps for creating concepts and designs
- Steps for testing and analyzing solutions
- - - Cycle repeats every 12 months

Mindsets mapped to design phases

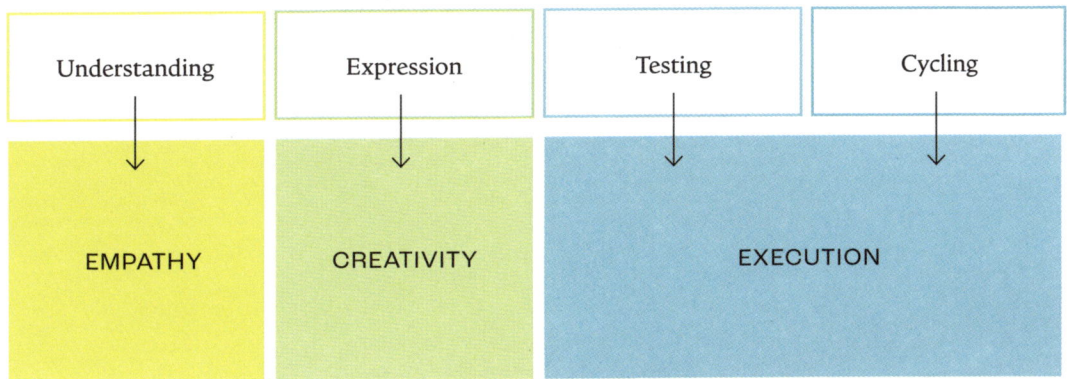

there is no standardized series of steps that will guarantee innovation and adoption. Again, it's the practice and constant improvement of the design process that determines the success of any creative endeavor.

Simplified, the design process can be summarized as stages of understanding, expression, testing or evaluation, and cycling or iteration. The testing phase tends to yield new information, which provides a new understanding of customers and stakeholders during the process. This then leads to new creative outputs that warrant further testing. Also note that the mindsets in these phases are all different. In an understanding phase, the designer must use an empathetic mindset. In the expression phases of the process, the designer relies on their creative mind. And in the testing and cycling phases, a critical mind is necessary. To be successful, a designer must use their whole mind, and approach their craft holistically. Whether or not you are empathetic, creative or inductive, logical or deductive, determines your success in each phase. When employed in a creative endeavor, are you making space for these phases of empathy, creativity, and critical thinking? Are you spending time evaluating your design concepts with those who are truly affected by them? What about iteration—allowing multiple cycles of designing and testing with your customers and stakeholders so as to uncover hidden human needs and evolve an idea from a mere suggestion to something deeply heartfelt?

As stated before, the neuroscience proves that you cannot blend these mindsets, and if your schooling or training has tended to focus on one particular frame of mind, then you owe it to yourself to broaden your design process by continuing to develop the mindsets you're familiar with, while also improving upon your weaknesses in others. The aim of this book is to provide lessons in developing these mindsets—empathetic, creative, and critical—which will help you generate resonance with consumers and unlock latent demand in the market.

An overview of design processes

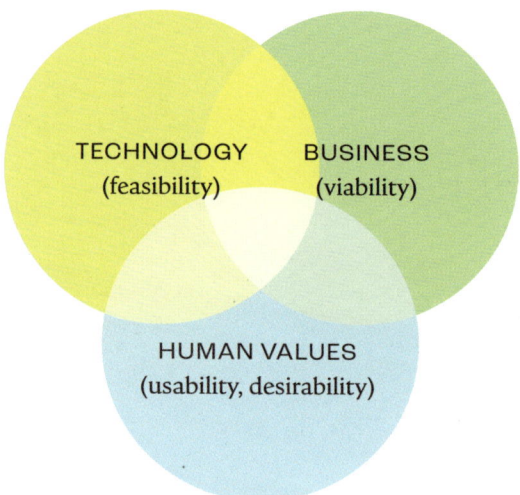

ABOVE: Design thinking uses a triple Venn diagram of business, technology, and human values.

Design is diverse. It spans a vast array of approaches, from larger-than-life celebrity designers who espouse a certain sense of style, to human-centered methods that look to real-world engagement as a source of inspiration. Design processes are shaped by the academy, the design industry, and educators researching the field. Industry practitioners also adapt and shape their own processes to fit both their work and their organization.

Having looked through both a theoretical and a real-world lens, let's examine a few more design processes out in the industry to become more acquainted with how organizations and businesses approach them. Many companies are happy to share their design processes with the public. However, not all of them have examined how their design process truly affects their business, sales, or customer resonance.

DESIGN THINKING

Design thinking is a school of thought that was devised at the Stanford d.school, framing the field on theory, practice, and disciplinary evolution. The methodology came about due to the need to "reinvent" the university around a model that used multidisciplinary collaboration and experiential learning across all departments to tackle real-world problems. Stanford had established an interdisciplinary degree in product design in 1958, and the d.school itself was founded in 2004 by Bernard Roth and David Kelley (who is interviewed on pp. 114–21).

This is a design philosophy founded at the intersection of three main tenets: technology (feasibility), business (viability), and human values (usability and desirability, which incorporate fields like the social sciences, art, and industrial design). Design thinking considers human values, or stakeholders (see p. 70), to be

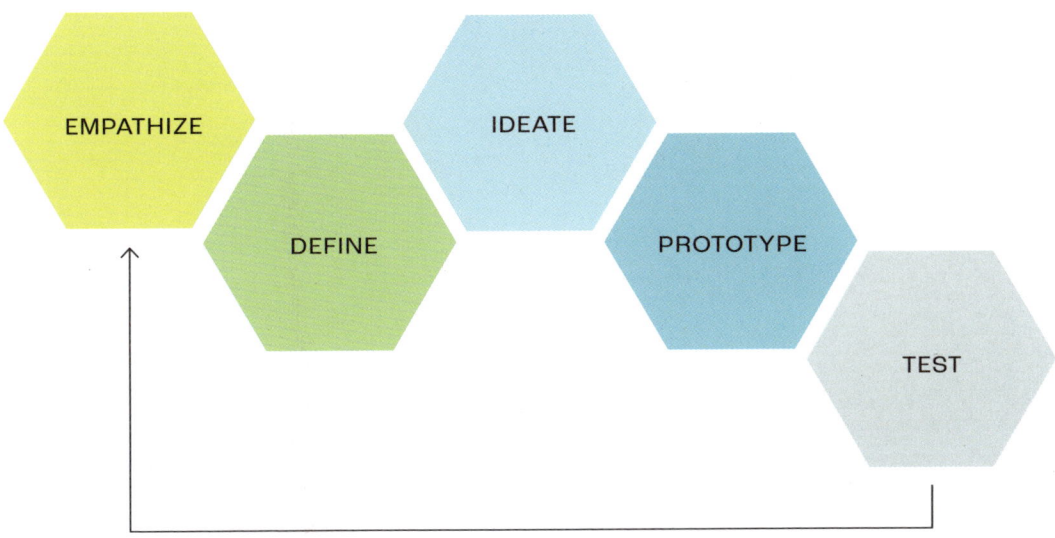

ABOVE: Design thinking's human-centered design process.

of equal importance to technology and business, which is unique among design methodologies.

HUMAN-CENTERED DESIGN

Design thinking has developed a design methodology with tools intended to engage people as a participatory element throughout the design process. It considers the triple Venn diagram as a starting point and introduces design tools in each phase of design. These phases are: empathize, define, ideate, prototype, and test. It's assumed that at the end of a test phase, a practitioner will have gained new understanding of the customer, and new empathy, and thus the process will repeat (cycle).

The term "human-centered design" (HCD) is often used synonymously with "design thinking." HCD was derived from the "user-centered" design theories espoused by Robert McKim and James Adams (at Stanford in the early 1960s and later), which treated the computer user as paramount to computer/interface design. These principles revolved around trying to "humanize" the computer interface, and make computing accessible—designed according to how humans wanted to relate to a computing machine, rather than conforming to the way the machine itself worked. A graphic user interface (GUI) involved a pointing device (mouse) and a display, and was more accessible to an operator than a formula translater (punch card using Fortran, a programming language), which was similar to typing line-by-line machine language code. HCD was expanded beyond the terminal user to include all humans involved (in an activity). It encompasses many methods that have been developed over decades of designing for others. Many of these, such as "brainstorming," have been appropriated by academia and industry to such a degree that the original intent has been lost to the cultural zeitgeist.

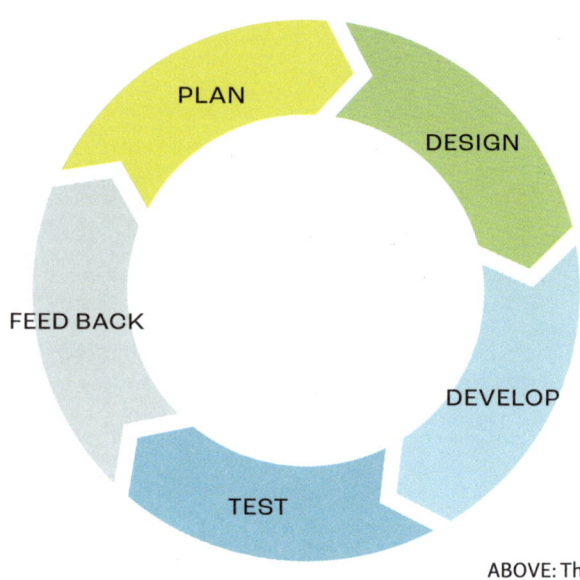

ABOVE: The agile methodology development process iterates around user testing.

AGILE METHODOLOGY

Agile was developed out of the computer science space (particularly through Google). Google's founders were PhD students from the Stanford computer science department and had been taught user-centered design approaches in their coursework. When they founded Google in 1998 their approach to software design contrasted with that of other companies, which tended to wait for two or three years before releasing new versions of software. As an example, the Gmail prototype was released to the public quickly, labeled as "forever beta," and feedback was solicited from users to help improve the software through iterative cycles. The innovation of Gmail at the time was the large storage space allowed to keep all of your email online on Google's servers, in addition to a text-search functionality that allowed you to quickly find a topic within your email threads, playing to the advantages of Google's search technology.

Much of the software industry has since followed this approach, releasing free or short-term trial versions in order to build a customer base quickly. These customers then provide feedback for refinement and suggest new features, which are quickly implemented and rolled out for future feedback.

Rather than being a gated, staged, internal development process carried out over years, the process is replaced by "sprints"—lasting a few weeks to a handful of months—that are quickly iterated and built upon. In this way, customer expectations can be quickly identified and validated. This methodology has often led to a model of software more as a subscription or service rather than discrete products.

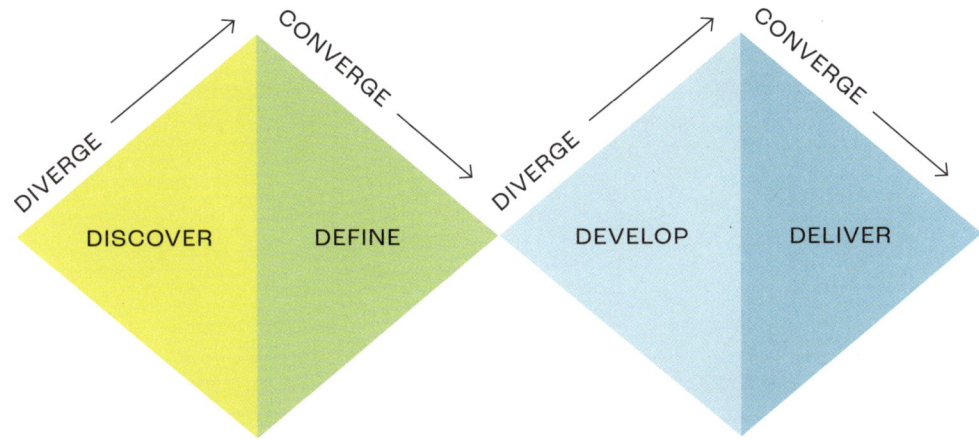

ABOVE: The Double Diamond design methodology alternates divergent and convergent thinking.

DOUBLE DIAMOND

The Double Diamond design process is a framework that was created in 2003 by the British Design Council, a strategic advisor for design incorporated by royal charter. This framework divides the design process into four phases: discover, define, develop, and deliver. One of the characteristics of the framework is that it calls for both divergent and convergent modes of thought, where divergent thought can be considered exploratory and creative, while convergent thought is more analytical and critical. The Double Diamond involves two rounds of these alternating mindsets—one for the definition of the problem and one for the creation of the solution.

The Double Diamond methodology has been incorporated in many areas of design. It can be found in consumer-product companies such as Kimberly-Clark and Proctor & Gamble.

Three "D" process: Note that the Double Diamond process has similarities to the Three "D" process used by European design firm frog (previously frog design), known for the first flat-screen TV design for Sony, and Windows Vista for Microsoft. The three phases are 1) discover: as a way of gathering insights into the problem, 2) design: incorporating both problem definition and the development of solutions, and 3) deliver: the implementation of the final solution.

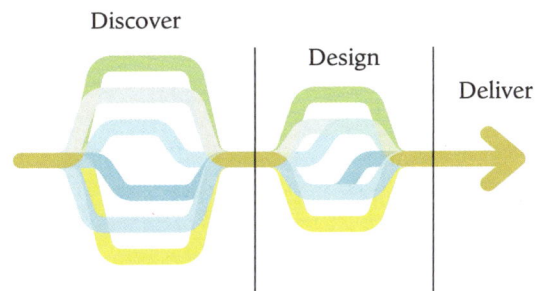

1.4 DESIGN PROCESS | 67

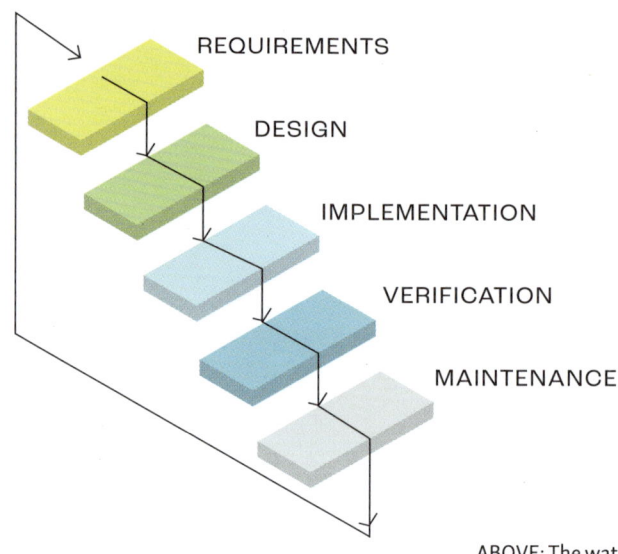

ABOVE: The waterfall design process uses gated phases for risk mitigation.

WATERFALL

Waterfall methodology is still practiced in more traditional, risk-averse companies, or where compartmentalization and a command-and-control structure are priorities. Large, conservative companies and government entities, like the US Department of Defense, will still use this development process. In these cases, strong, explicit documentation of each phase minimizes training for workers new to a position, as the process tasks are followed rigorously, with little deviation, which provides a certain level of consistency between milestones. Timing and costs for a project are also easier to estimate.

Compartmentalization is a characteristic trait of this process, for when secrecy between specific groups within an organization is important, or when a product/design only requires incremental improvement since customers' needs are well understood and change very slowly.

The disadvantages of this methodology, however, are a lack of adaptability and responsiveness to changing customer needs. The rigidity of the process and the lack of engagement with customers outside the first phase means that new requirements or changes halfway through can't really be addressed. Since all phases are sequential and independent, a delay in one phase will cascade downwards. Waterfall doesn't put the customer or their requirements at the center of the process, nor is there any ability to flow information backwards through the teams (from maintenance back up to requirements) or between design teams at the cycle point, which are inherent weaknesses in the process. Also, due to the siloed nature of the process (waterfall is also known as "over-the-wall [or cubicle]" design), each silo/organization is not required to follow or communicate with the previous group. In industry, each step here is represented by an independent department:

ABOVE: Students at Middle East Technical University in Ankara, Turkey. Collaborative exploration requires cooperation but it also takes advantage of differing points of view to solve a problem.

requirements = marketing; design is self-explanatory; implementation = engineering and manufacturing; verification = quality assurance; and maintenance = customer support. These functions are not performed by the same people, so communication of intent and adaptability to change is very limited.

ADVICE ON DESIGN PROCESS

Design processes are varied, born from organizations, and practiced individually and among groups. As you gain experience navigating your own projects, you'll begin to develop an individual design process that's uniquely yours, informed by how you work, the tools you prefer to use, and your intuition. Due to the tendency for groupthink to arise in an organization, a particular work environment may not always have the process that suits you best. An organization may also be in a stage of reinvention itself. Finding ways in which you can practice your individual process within a large organization while still conforming to its development methodology is half the battle of professional practice.

The other half is self-discovery. Be critical of your own process. After a project or phase, gauge what went right and what went wrong. Given the opportunity, would you approach it differently? What could have been improved? Might it even be necessary to repeat the phase, or start over completely? As you hone your individual design process you'll identify potential missteps earlier and more easily. The main advice I can give is that you remain responsive to your process: aware of it, adaptable to change, and open to improvement.

Tool | Stakeholder Matrix

Now that we've covered design processes, how might you go about creating and improving your own process? The tools provided over the next few pages will help you to engage with people throughout your design process and give you ideas for how you might want to work with them. This is an opportunity to expand upon your empathetic mindset.

In *Participation and Social Assessment: Tools and Techniques*, Jennifer Rietbergen-McCracken and Deepa Narayan describe two methods to help plan engagement with others throughout the design process—a stakeholder matrix and a participation strategy. We'll begin by looking at the stakeholder matrix, and the participation strategy will follow on pages 74 to 75.

Who or what are stakeholders? These are individuals, groups, or institutions that may either: a) be affected by the proposed design or b) influence the outcome of your design, by hindering or helping to facilitate its development. Some common stakeholders include beneficiaries (people who benefit from your intervention), competitors, funders or finananciers, colleagues, regulators, governments, and pressure groups (social activists), just to name a few. You may be able to think of others.

Before plotting your matrix, you'll need to do some preparatory work, assembling certain information about your key stakeholders, and then ranking them in terms of influence and importance.

IDENTIFY YOUR KEY STAKEHOLDERS

Download the template opposite (using the QR code on p. 72), then start by completing the first column. Identify your potential beneficiaries, then consider who might be adversely impacted by your design. Once your supporters and opponents have been identified, you'll begin to recognize the relationships that exist among them. (In the next step we'll also ascertain the vulnerability of each stakeholder group.)

ASSESS STAKEHOLDER INTERESTS AND IMPACT

In order to complete the next two columns, think about what the stakeholders' expectations might be. What benefits are there for each? What resources might each stakeholder be able and willing to contribute? What stakeholder interests conflict with your project goals? Fill in their interests and impact accordingly. Pay attention, again, to the relationships between stakeholders; sometimes, relationships you haven't anticipated will come into play. For example, the relationship between a manufacturer and a supplier is mutually beneficial. However, if you have two suppliers providing the same part, the relationship between them becomes competitive, and could become adversarial. From your manufacturer's perspective, on the other hand, this might be a desirable scenario, potentially allowing them to pit one supplier against the other to negotiate the best price.

Having completed these two columns, you'll have the information you need to be able to assign a value in the final two columns.

Stakeholder	Interests	Impact	Influence	Importance

ASSESS STAKEHOLDER INFLUENCE AND IMPORTANCE

You'll now assess each stakeholder's level of influence and importance. Let's first define these concepts:

- **Influence:** refers to the power that a stakeholder has over your project. It's their ability to affect the objectives and direction of an intervention or organization. They can help or hinder your project's implementation. This control may come from a stakeholder's status or power, or from informal connections with regulators or social-activist groups.

- **Importance:** is the degree to which a stakeholder is affected by an intervention or organization. If your project is attempting to address their needs, and will greatly affect their life, their interests will match your project's objectives.

Consider each stakeholder's power and status (political, social, and economic), degree of organization, control of strategic resources, informal influence, and power relations with other stakeholders. On a simple numerical scale (1 to 10) rank the influence and importance of each stakeholder.

The 2x2 matrix

Using their numerical rankings in terms of influence and importance, map your shareholders on a 2x2 matrix, as shown opposite. This tool provides a way to understand the power dynamic of each stakeholder. Let's take a look at each quadrant in turn, based on the completed sample shown bottom right.

Q1: HIGH IMPORTANCE/HIGH INFLUENCE
Everyone wants to get this quadrant right, so these stakeholders require close involvement. They are very important to your project's adoption but they also have the potential to derail it. These people should be consistently engaged. An example, if you were developing a baby stroller, would be the parent. They are the customer for your device, and its design will greatly affect their life.

Q2: LOW IMPORTANCE/HIGH INFLUENCE
These people are not the target for your project but they could affect its outcome. Groups such as regulatory bodies will have no personal interest in your project's success but if they're not kept informed and you run foul of their rules, they'll prevent your project from happening. In this case, a consumer-product safety board will govern your baby stroller's adherence to standards. They are not personally invested in or impacted by your product, but you must abide by their safety regulations.

Q3: LOW IMPORTANCE/LOW INFLUENCE
You can think of this group as having the same interest in your project as the general public, so no special treatment is required. You can keep them informed using the same information-sharing strategy that you'd use for a general audience, letting them know that a new baby stroller is in development.

Q4: HIGH IMPORTANCE/LOW INFLUENCE
This is the quadrant that designers can easily miss, so it's one to look out for, particularly as this is the most vulnerable of all the groups. The fact that these stakeholders have little say in your product's development yet are greatly impacted by it means that special attention and effort are required, ensuring that their needs are met and that their voice and participation in the development are taken into account. In this case, the most vulnerable stakeholder is the baby in the stroller. She has no power over the product's purchase, and the design is inflicted upon her.

Download the templates and post your matrix by using these QR codes.

2x2 matrix

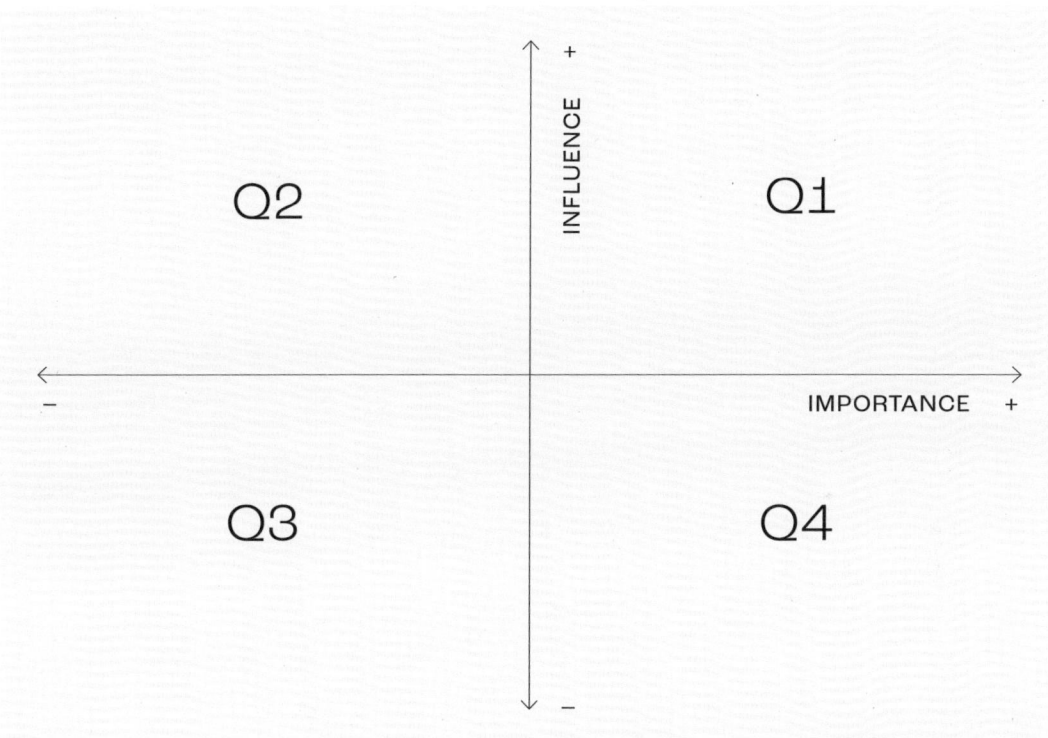

Sample

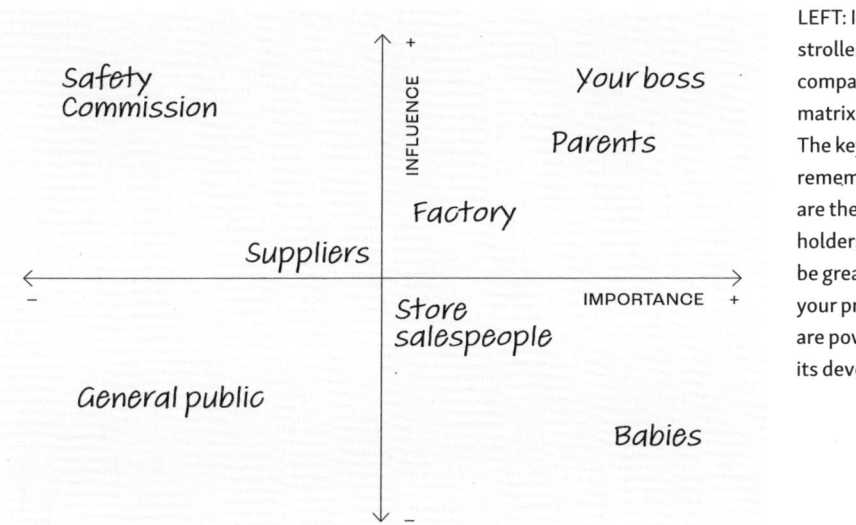

LEFT: If you were designing a stroller for a baby-products company, your stakeholder matrix might look like this. The key quadrant to remember is Q4. These are the vulnerable stakeholders—those who will be greatly affected by your product but who are powerless over its development.

Tool | Participation Strategy

You'll now draw on the information gathered for your stakeholder matrix for a different tool—a participation strategy, which will chart your design process over time.

This is a map of your design process in relation to each stakeholder, detailing how you will engage with them. You will plan involvement according to the interests, importance, and influence of each group (particular efforts will be needed to involve important stakeholders who lack influence), then determine appropriate forms of participation throughout each cycle (evaluation period) of the project.

Some questions to ask when identifying design specifications and human requirements:

- What are your underlying assumptions?
- How malleable is the task/assumption? In other words, what may change over time?
- Who are the main influencers (stakeholders) behind each task?
- What level of evidence/need exists for each stakeholder, or is needed for the task to advance (importance)?
- Are there any other outside factors?

Participation-strategy timeline

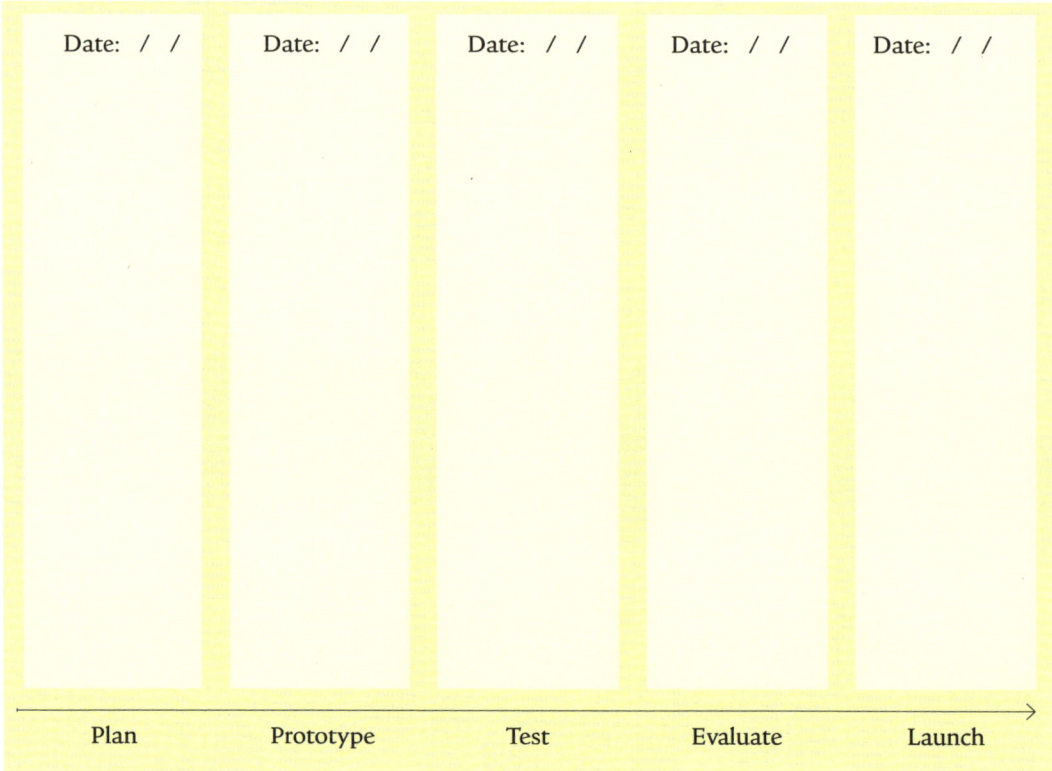

74 | CHAPTER 1: LAYING A FOUNDATION

KEY TIPS

Remember that this is a living document and that, in addition to the stakeholders who'll be present throughout, others will come in and out of your process over time, depending on their relationship to your project and their placement in the 2x2 matrix. You will go through several cycles of the design process, so think of this document as something that's used repeatedly, over each iteration.

Make sure that you choose the appropriate form of engagement. For a Q2 stakeholder, a phone call reviewing particular regulations against your product might be sufficient. However, for a vulnerable Q4 stakeholder, would you be sure that with just a phone call you'd be giving that group sufficient voice—empathizing with the issue at hand, as well as providing sufficient involvement with the creation and evaluation of the design? And how often would a Q1 stakeholder (high importance and influence) need to be involved, and in what manner? Would it be a weekly call or review? These are all choices you can make as a designer/team, but identifying the most appropriate form of engagement during each design phase will help move your process along efficiently and effectively.

Sample

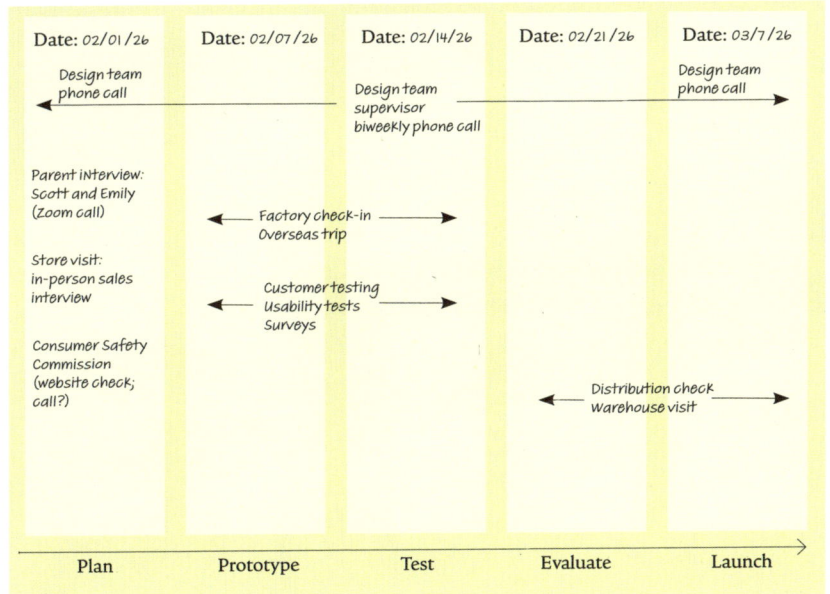

LEFT: Using the baby-stroller example again, the participation-strategy tool maps stakeholders to a timeline, ensuring that you engage with them throughout your project. It's a living document that guides you through your design process.

Download the template and post your participation strategy by using these QR codes.

1.5 | Case Studies

Let's take a look at a couple of case studies of products and their acceptance in the market, thinking about how the design process helped shape the outcome of the design, for better or for worse.

HUMANITARIAN DESIGN
One Laptop Per Child (Africa) versus the microfinance system (India)

While I always applaud those creative individuals in the humanitarian design space, many of these efforts don't succeed, and this is usually due to bad design process. Launching in 2005, the One Laptop Per Child (OLPC) program was an initiative from Seymour Papert and Nicholas Negroponte of MIT's Media Lab, who wanted to lower the price of a computer drastically, making it more accessible in developing countries. The goal was to reduce the price of a laptop from around $1,400 (in 2005) to $100. In their first concept study, they used a well-known Western designer—Yves Behar of Fuseproject—to conceptualize the laptop, the XO-1, which of course brought notoriety and media attention.

Though OLPC was founded on a researched principle that providing computing technology earlier in a child's life improves their education, several issues arose with the adoption of this design. The program conceptualized the laptop first, without regard to context. Sales of the XO-1 were done in batches of 1 million computers, which was later lowered to 250,000. Without any evidence of benefit besides a slick concept model, asking a developing country to pay $100 million would seem like a risky proposition. That amount of humanitarian aid could go to a lot of other sources with proven areas of benefit. The software was also a debate between Windows XP (for $3 each seat) versus open source, in which the philosophy of who owns the operating system was not considered.

LEFT: The OLPC XO-1 designed by Yves Behar was a rugged laptop concept with a $100 price tag, aimed at educating children in developing countries.

When laptops were delivered to their destination, students were given complete permission over the device and they uninstalled the software that they were supposed to learn. Though the design was rugged, there was a lack of service, support, or spare parts because the infrastructure around supporting the design had not been well thought through.

Finally, and most importantly, there was a fundamental assumption that merely lowering the price of a technology used by a developed nation was something that a developing country wanted. Children in each locale had different customs, traditions, and methods for learning. There was no emphasis on localizing the offering or seeking input from the students, educators, schools, and local administrators toward the design of the system, which caused distrust in adoption. Instead of designing a slick-looking laptop, engaging iteratively with the culture through rounds of making may have helped identify what was actually needed in a laptop—or if a computer was even the right technology to offer in the first place.

We can contrast this with the development of the microfinance system in India. Instead of starting with a technology and looking for a use case (known as "hammer looking for a nail" syndrome), start with the culture and identify the frame you'd like to change by engaging with the people directly. Microfinance refers generally to the provision of financial services, including small-value loans, to households, small businesses, and entrepreneurs who lack access to formal banking services. By observing the lack of financial inclusion for women in India in rural areas and households at or below the poverty line, and empathizing with the constituency, other banking models were established. These included joint-liability groups (JLG), consisting of four to ten people working together as a team to apply for bank loans, and self-help group (SHG) bank-linkage programs—a group of microentrepreneurs (perhaps a store owner or a number of people with the same socioecomonic status) getting together voluntarily to contribute to a common fund that is borrowed against. This system leverages the appropriate technology (cell phones and Paypal-type apps) to replace the logistics of more established banks. The microfinance system addresses the needs of the people first, then develops with the constituency in tandem, refining the offerings over time. This empathy, create, test type of practice is repeated and refined constantly.

COMMERCIAL DESIGN
Xbox 360 (Microsoft, 2005) versus Wii (Nintendo, 2006)

Video-game consoles are an interesting product design and business model, where design process can really illuminate both context and empathy concerns. As a console maker, you're running a platform business (versus an end-to-end, pipeline business). It's a catch-22 situation. As a platform, you have to provide a gaming experience that customers want to participate in, but on the other side, you have to sell enough consoles, and have enough of an installed base of users to entice software developers to create games for the platform.

Prior to 2005, Microsoft's Xbox and Sony's PlayStation 2 (PS2) were the dominant consoles on the market. For their follow-up consoles—the Xbox 360 and PS3 respectively—they both

their own internally made games as well as the games that they licensed for the platform.

Nintendo, on the other hand, took a different tack. The company was in a tough financial position because their last game console, the GameCube, had suffered low sales compared to the Xbox and PS2. Though advanced for Nintendo, the GameCube was easily outpaced by other consoles in terms of graphics. Nintendo knew that they wouldn't be able to beat Sony or Microsoft at their type of "hardcore" gaming, so for their next console they decided to look at gaming through the eyes of a different customer.

What simplifying assumption had the incumbents made about video games? That hardcore games demanded cutting-edge realism. However, why do people play games in the first place? To have fun together. Thus, Nintendo sought to redefine fun through the eyes of a family playing together—parents or grandparents playing with young kids. To appeal to this wider spectrum, the company introduced a different way to play their games. The control interface was a motion-based controller instead of a joystick and thumb buttons. This didn't depend on quick reflexes and twitch muscles but on much larger swings of the arms and gestures that younger children and elderly players, who both lack full hand–eye fine-motor coordination, could play more easily. This control type didn't require the sophisticated computing hardware that its rivals did either, so each console, although less powerful, actually turned a profit too. Nintendo also bundled their own first game, Wii Sports, so that you could begin playing right away. The name of that console, the Wii, was pronounced *wheee!* (fun), or *we* (together).

followed the model that had earned them strong sales in the first place. Firstly, they focused on improving the graphics and processing power of their games, making them more realistic than ever before. More realistic games were what they believed the "hardcore gamer" market (primarily male teenagers) demanded. For the Xbox, the signature game series was Halo, a first-person shooting game that relied on quick hand–eye coordination and twitch reflexes. This was the game that their market liked playing. In addition, because of the drive to one-up each other, both companies' consoles employed cutting-edge processors and graphics chips (at greater expense for Sony for the PS3). Both companies employed a loss-leader model, where they sold an advanced gaming console at a loss but would then make up that money through royalties on

Even the "ii" was intentional—it represented two people standing next to each other.

How did the consoles fare? While the Xbox 360 and the PS3 split the hardcore market, the Wii expanded upon that market at either end, selling to parents, the elderly, and younger gamers (children, not teenagers). What did that mean in sales and profit? As of 2024, the Xbox 360 had sold 85.7 million units globally at a loss, and Nintendo had sold 101.6 million units at a profit. As to games, in 2007, the Xbox 360 had three games selling in the top 10 list: Halo 3, Call of Duty 4, and Assassin's Creed, for combined sales of 9.73 million units. During the same period, Nintendo also had three games in the top 10: Wii Play, Super Mario Galaxy, and Mario Party 8, for a total of 8.46 million units. From a revenue standpoint, Nintendo had built a more successful product.

Looking back at this case study, what simplifying assumptions did Microsoft and Sony make about their customers? In their place, how might engaging with other types of customers interested in gaming have shaped what you would have done differently? How might testing with your users have affected this outcome?

OPPOSITE: The XBox 360 was Microsoft's follow-up to the original Xbox.

LEFT: The Nintendo Wii introduced a new type of motion-controlled game play and a larger, more inclusive customer market.

1.5 CASE STUDIES | 79

2
DESIGN BEHAVIORS
Design as a Whole-Life Activity

DESIGN IS NOT A DISCIPLINE, IT'S A LIFESTYLE. WITH PRACTICE, WE LEARN BY DOING AND MAKING. EXPERIENCE LIFE. IMPACT OTHERS.

2.1 | Behavioral Approaches
2.2 | Design Empathy
2.3 | Contextual Awareness
   Tool: AEIOU
2.4 | Creativity and Craft
   Tool: The Elements of Art
2.5 | Rapid Iteration
2.6 | Entrepreneurial Sustainability
   Tool: Hierarchy and Principles of Design

2.1 | Behavioral Approaches

We can see from the previous chapter's investigation of the design process that the more adeptly a designer navigates through shifts in mindsets, utilizing different areas of the brain at the appropriate stages of the process, the more relevant and effective the outcome is for the intended audience.

Whether you're in a creative, inductive mindset or a critical, analytical one, cultivating suitable behavioral approaches will expand your experience base and strengthen your "creative musculature." Behaviors are learned over time, through repetition, cultivation, and reflection. And the same is true when applied to design tasks—specific approaches can be learned and refined through practice.

This chapter will cover five different design behaviors: design empathy, contextual awareness, creativity and craft, rapid iteration, and entrepeneurial sustainability. In my experience, regardless of industry or final product, refining and practicing these behaviors will improve your output and your sensitivity to the needs of your audience. We'll therefore talk briefly about each behavior and how it can shape your process. Then in the following two chapters, we'll take deeper dives into design empathy and contextual awareness in particular, providing: 1) concepts for exploration and 2) tools and methods for putting them into practice.

OPPOSITE: With practice, these five design behaviors will become habit. When ingrained in your work process, they'll enable you to create meaningful solutions faster—solutions that will resonate with your target market.

DESIGN EMPATHY

An open mind, tolerance for others, and perspective taking are empathetic skills used by the 21st-century designer. Seek first to understand others who are different from you, rather than judge or categorize them. The ability to "method act" someone else's life (see pp. 140–43), rather than dictate their behavior, will speed the adoption of the design.

CONTEXTUAL AWARENESS

Today's practitioners are required to focus attention on context and situation. Current trends in meta-thinking engage the modern designer contemplating not what is known or seen, but the implicit, hidden forces at work. This state of inquisitive curiosity spurs innovative solutions that disrupt markets.

CREATIVITY AND CRAFT

Bring any idea, whether it be a product, environment, service, performance, policy, or business, to the point where it can be realized and tested against an audience. Flexible prototyping skills that match the prototype's fidelity and resolution are appropriate for this stage of development.

RAPID ITERATION

Additive manufacturing, low-fidelity prototyping, agile development, and flexible manufacturing reveal a trend—ever quicker cycles between the expression of an idea, and feedback from an audience. You gain more from multiple cycles made with feedback than from a single, high-resolution solution.

ENTREPRENEURIAL SUSTAINABILITY

For a design intervention to endure, the business model must be sustainable. Value, benefits, resources, and costs must be in perfect balance. All the great designers have been keenly aware of this alignment. As Thomas Edison said, "Anything that won't sell, I don't want to invent. Its sale is proof of utility, and utility is success."

2.2 | Design Empathy

Empathy is more than compassion. One can share a concern about someone's well-being, but that's not design empathy. The latter involves the empathetic skills and mindsets you can use to "get into character"—in other words, understanding others, both cognitively and emotionally, through their experiences and then being moved to act. In order to "method act" someone's life, you need to understand not only their outlook and perspective, but also your own views in relation to theirs. If you encounter someone with a sprained ankle, for example, there's a difference between feeling sorry for that person's situation, and actually experiencing the discomfort of a sprain, the frustration of limited mobility, and the impatience for a recovery that is slow in coming. Design empathy encompasses empathy in all its forms. The psychologists Daniel Goleman and Paul Ekman have defined three different types of empathy.

The first, cognitive empathy, is the ability to understand another's perspective. This involves simply knowing how someone else feels or what they might be thinking. Sometimes referred to as "perspective taking," cognitive empathy is the first step toward understanding the world through the eyes of another.

Emotional empathy is the ability to channel the emotional state of a person, as if their emotions were contagious. Using the sprained-ankle example again, emotional empathy would be your ability to channel pain, frustration, and impatience, and to the same degree as the injured person. For example, if you're particularly sensitive to pain, but the person you observe habitually sprains their ankle due to their occupation, can you channel the level of frustration or pain that they'd feel? They have most likely simply become inured. They still experience the pain, but perhaps not at the same level that you would, being sensitive to pain and it being a first-time experience. Can you imagine the feeling of a repetitive injury and how that pain may be dulled—still present but more of a nuisance in the back of your mind than an acute, front-of-mind awareness?

The last type of empathy, which draws on both cognitive and emotional empathy, is empathetic or compassionate concern. Not only can you understand someone's situation and feel along with them, but you are also spontaneously moved to help, if needed. Our ability to channel our empathetic understanding into action that can address the situation is a critical part of framing a design problem.

Just as creativity and critical thinking utilize specific areas of the brain, empathy requires the use of the empathetic centers within our brain. The Max Planck Institute for Human Cognitive and Brain Sciences discovered that the right supramarginal gyrus (SMG) area of the brain is responsible for compassion and regulating empathy—distinguishing our own emotional state from that of others. Further, the researchers concluded that compassion and empathy can be acquired and developed. Quiet meditation, practicing acts of kindness, and understanding different walks of life are all ways to develop empathetic mindfulness.

Design empathy, then, is the combination of these three types of empathy in order to better understand the audience of your product but with regard to your own perspective. This means understanding your own life situation and how that has contributed to your outlook on the world. Part of the journey to empathetic skill is an assessment of our outlook on life and the sources of influence (education, power, finances, and so on) that have shaped it. Understanding others' perspectives, while also understanding where your own perspective may influence or unduly affect others, is a key aspect of learning and mastering this behavior.

Luckily, a design empathy mindset, like all design behaviors, can be practiced and improved. Whether it's the method-acting techniques that a performer would use to internalize a character's motivation, or the neutrality employed by a social scientist to engage others in conversation, there are ways to help unlock an unbiased understanding of others, and some of these will be explored in the following chapter.

The three types of empathy

COMPASSIONATE
CONCERN

COGNITIVE EMPATHY

The ability to understand another's perspective. Simply knowing how another person feels or what they might be thinking is cognitive empathy.

EMOTIONAL EMPATHY

Emotional empathy is the ability to channel the emotional state of another, as if their emotions were contagious.

2.3 | Contextual Awareness

ABOVE: In the inner circle, a lamp sits on a table. This is then expanded out to the living room, the house, and the city. Beginning with the lamp, can you imagine expanding out to further and further contexts?

Contextual awareness is a designer's secret sixth sense. Naïve creators often mistake the problems they face and therefore propose solutions that don't actually address the problem or miss the mark, despite noble intentions. This is almost always because they don't understand the context and environment in which their design solution is to be implemented.

Context involves an understanding of the myriad factors surrounding a design's implementation. On the one hand, it is understanding the levers of business and production to match a design's mass production to the available demand in the market. On the other hand, it is a sensitivity to the use case, culture, and environmental factors surrounding the adoption of the product. Just as empathy is a behavior that seeks to understand an individual and their experiences from their own point of view, contextual awareness provides the methods to understand a culture or tribe of people, and the mores and values that establish their collective outlook on the world. Contextual awareness also provides insights into how those values may overlap, conflict, and contradict each other. Further, it provides the wherewithal to position a design intervention at the intersection of the conflict to mediate a better outcome.

What's interesting about context is that so many of the factors that drive context in design situations are neither found in textbooks nor are they easily recorded. Because of the elusive nature of context, practitioners often mistake what they think is the correct way to implement change in the world. Context is informed and inspired by abstract factors, which are often implicit, or hidden. They exist within the culture, but are not explicitly stated. Implicit factors, cultural norms that are readily understood and accepted, but never explicitly stated, are prevalent throughout societies of all kinds. If ignored, they can be a serious impediment to a market's adoption of a design. Contextual awareness revolves around a sensitivity to these factors, and an understanding of how to identify, organize, and prioritize them to position a product for maximum market penetration.

Contextual awareness itself—often a behavior practiced when defining a problem—alternates between creative thinking and critical thinking but it's also abstract(ed) by nature. This design abstraction looks beyond the concrete things that we can observe and record and considers the implicit, hidden factors behind them, both creatively and critically. We are both creating hypotheses on the working condition of the product and testing those theories against people's views in reality, and this is known as "research synthesis." In this way, our thinking must consider the abstract—imagining qualities that are greater than what is seen, as well as thinking analytically about how those factors came to be.

Tool | AEIOU

To use observation as a design tool, you go out and observe an activity or a scene in which your design might be used. It's important to observe in the natural environment, out in the world, rather than trying to replicate an experience in a lab or a contrived simulation.

There's a difference between being an active observer and a passive participant. Active observation entails seeing something with "new eyes," as if you've never seen it before. If everything is new to you, you are actively interested in watching and learning more about it. Don't rely on prior understanding of a situation or on patterns that you've been aware of previously. You need to have the mentality of an inquisitive child, or of someone who just landed on the planet yesterday—a viewpoint that's called a "beginner's mindset."

We can observe any setting where people who (by playing different roles) create an experience, perform a task, deliver a service, or solve a problem for others. It's important to build upon your observational skills. We'll cover interviewing and engaging with people in Chapter 3, but actively watching, taking it all in, noticing all the little details, and becoming highly visually observant are all fundamental skills for design.

When conducting an observation, you can record what you see using readily available tools—a notebook for handwritten notes, and a cell phone with photo, video, and audio recording functions. You can take pictures and record video and audio to capture your observations for later reference, and once you've begun developing insights about your audience, you'll find that you often review your notes.

A general approach for observations is to start taking notes first. When everything is new—when you're approaching it with a beginner's mindset—you might feel overwhelmed with the amount of information you need to absorb. Notes will give you an initial foothold.

As with a marathon, the first part of the process might be positive and interesting, but eventually your eyes will get acclimated to the environment and routines of the people around you, and you'll sink into the tedium of the race, which usually means fatigue, boredom, and frustration. This is not the time to quit. As for the marathon runner, this part is a necessary evil; you'll have to break through the runner's wall and catch your second wind. This is, in fact, the time to keep watching because it's usually after this point that you'll experience an "Aha!" moment, struck by something unexpected. This is known as an insight—a non-obvious, interpretive statement that reveals an unspoken belief or action that's supported by your observations. When studying a dog park, for example, an insight might be: "When dog owners and their dogs play together, owners who resemble their dogs tend to socialize together and have purebred dogs, while owners who don't resemble their dogs tend to socialize with the owners of mixed-breed dogs."

OPPOSITE: Feel free to experiment with your notetaking. Do you like to use colored highlighters to distinguish different groups? Do you like to take photographs, use a scrapbook, jot down annotations in the margins? You'll soon find a style of recording notes that works for you. If you make the activity playful and creative, you'll be inspired to do it more often.

Rick Robinson, the founder of E-Lab, devised a great observation framework that he named AEIOU—a mnemonic that acts as a handy reminder of the key elements to include. When conducting an observation, be sure to record the following:

ACTIVITIES
What are the actions and behaviors that you can see taking place? These can be step details or overall descriptions (ordering a coffee at a coffee shop OR placing an order, paying for coffee, and so on).

ENVIRONMENTS
Describe the venue or overall setting that you're in, which will help to provide useful context.

INTERACTIONS
Record basic interactions you see taking place—between people, between people and objects, and even between objects themselves. Then create a timeline to help order those interactions.

OBJECTS
Note any objects, whether natural, man-made, or both. You may see patterns in what exists, where, and why. Material culture uses descriptions of objects to interpret the behaviors, values, and meaning structures of the people who use them.

USERS
Who are the people that you see in your observation? What are their names and job titles? What roles do they play in their community or social network (for example, fireman, sister, bully …)?

TOOL: AEIOU | 89

2.4 | Creativity and Craft

Creativity and craft are intrinsically linked. When approaching a product, there'll be a time when identifying issues and coming up with a maximum number of ideas to solve a problem will determine how much time you have to present or make each idea. In this way, the craft(smanship) of making is related to the level of directness with which you want to communicate. At other times, you'll be comparing different executions of the same idea, and judging individual characteristics or increasing the quality of production. In this case, the realism of the build and its accuracy to the desired quality will be much higher.

The sophistication of a presentation is dependent on where you are in the timeline of a project, your ability to craft things, your internal team's expectations, and the type of feedback you wish to gain from your target audience. Your ability to come up with ideas (ideation) and the expression and execution of those ideas (prototyping) through your ability to fabricate them is key. Craft matters, but matching these levels of craft to the right stage in your design process involves an understanding of communication, intent, and the skills that best leverage your creative output.

This design behavior is critical for managing resources within a project. The sophistication of your craft, and the resolution of your idea—its effectiveness at visual communication—can be considered its fidelity. A low-fidelity execution of an idea invites discussion and interpretation because its depiction is slightly vague on purpose, while a highly executed version of the same idea is used for persuasion: to sell the idea. A clear, purposeful communication of the design's intent is the objective of the pitch.

Creativity, as previously discussed, comes from the rapid switching between different networks in the brain. From the inductive, intuitive mindset (unencumbered by doubt, reason, or self-critique) used to generate ideas,

BELOW: Think of all the ways you can represent an apple. How does each image contribute something unique to its representation?

to a quick pivot—best thought of as abstract comparison—this switching weaves multiple threads into a new patchwork of ideas. Taken together, this creative mindset (inductive + abstract comparison) develops themes and allows contrary ideas to emerge, and the abstract collision of these disparate ideas generates new inspirations and new possibilities. Ideas can be both similar (sequential thought) and diverse (lateral thought). When carried out in a team setting, a culture of creativity is open to new and contrary ideas, and it produces a space in the project for wild and crazy proposals, collaboration, multiple perspectives, and improvisation. Creative and innovative cultures are those that are extremely receptive to a diversity of perspectives, and create spaces that invite others to contribute.

Creativity and craft leverage the creative mindsets that allow our inductive reasoning to imagine new possibilities and jumping-off points from existing points. They also leverage the experience and recollection of previous attempts at building in such a way that it provides insight into fabrication. Thinking through a fabrication process can help you understand how constituent parts might come together, and how that might communicate different intents. It can also provide insights into which aspects of your models will need improvement or reinvention. In this way, creative craft is both a creative and an experiential phase—you learn by doing or making. In any creative profession, practicing a craft and pushing your ability to execute under different circumstances not only affords you new knowledge, it also boosts confidence in your ability to imagine and build.

Tool | The Elements of Art

In my line of work, I'll often find myself next to a whiteboard in a business meeting or a working group, being asked to justify the importance of aesthetics for a product design relative to some other technical factor that's more easily quantified.

It's a difficult conversation to have—comparing disparate qualities, where one characteristic is easily measured (using numbers) while the other must be experienced to be understood.

Art and design require a human audience; they need to be engaged with in order to be truly appreciated. However, descriptions of more intuitive approaches—which are very familar to those who have an art-and-design background—can seem somewhat nebulous to those who specialize in data and analytics. In professional situations, the public and the client will not be privy to your entire design process; they'll only be

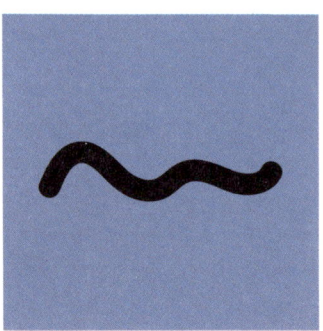  

LINE
A line is a simple concept to understand, but it can be a very complex idea in practice. It can be considered in two ways: 1) a mark made with a pen or brush or 2) the intersection created when two shapes or objects meet. Lines have character: the way they're drawn or present themselves has a personality. Horizontal lines connote struture or calm, curved lines are soft and organic, and zigzag lines suggest movement.

SHAPE
A shape is a self-contained, defined area, which is made using lines or edges. A positive shape in a design automatically creates a corresponding negative shape. In a given work, there's the object that you draw (the figure, which is positive) and the background or environment that surrounds it (the ground—negative). Shapes also have character—geometric, organic, and so on.

FORM
Form refers to three-dimensional shapes that have length, width, and depth. We know something has a form because it reacts to light and casts a shadow. Form can be appreciated as emerging from a surface (bas relief) or fully in the round.

shown an end result (a production-ready product) that will belie the amount of thought, understanding of process, and decision-making that went into its creation.

One approach that's helped me to explain the craft of design is to provide a very quick lesson on visual hierarchy and Gestalt theory (covered in more detail on pp. 110–11), otherwise known as aesthetics. Since, as we've established, the design fields straddle both intuitive and rational approaches, it's very helpful to establish a basis for why working visually trumps working verbally. To make this effective, we first need to introduce a few of the tools that designers routinely use in order to convey their ideas. Six of the most common of these—the elements of art—are line, shape, form, space, color, and texture.

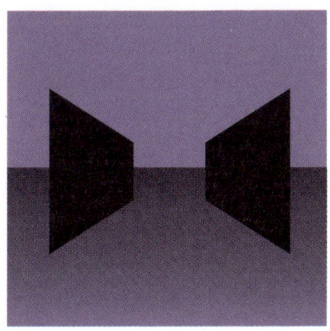

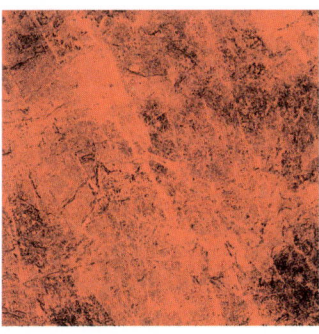

SPACE
Space is defined and determined by shapes and forms. Positive space is where shapes and forms exist; negative space is the empty space around those shapes and forms. Think about how forms and shapes may be compiled and arranged to arrive at a striking composition.

COLOR
Color is produced when light strikes an object and reflects back in our eyes. A color has three main characteristics:
Hue: Where the color is positioned on the color wheel. This is the color's name: red, blue, and so on.
Value: The color's lightness or darkness—how close to black or white it is.
Saturation: The intensity, or level of chroma. The more gray a color has, the less chroma.

TEXTURE
This refers to the surface quality or "feel" of an object—smooth, rough, soft, and so on. Textures may be actual (felt with touch—tactile) or implied (suggested by the way an artist has created the work of art, through the use of other elements). For example, a drawing may "feel" sharp or jagged, due to the use of bold zigzag lines and forms positioned close together.

Visual hierarchy

Let's now look at how these six elements of art can be used to create visual hierarchy. Industrial designers use these elements to design the interfaces of products. We can start with shape. What's the largest circle on a washing machine? And what's the next largest? The washer below uses a consistent shape (a circle) to lead your eye: The size of the shape suggests the order of engagement. One of the principles of Gestalt theory is "similarity," according to which all shapes that are similar tend to be perceived as related. For this product, the use of a circle as the control that you use in order to operate the machine is easily understood.

A good washing-machine design is so intuitive that you don't need instructions to use it; it simply functions as you'd expect it to. The largest circle (the door) is the first object that you perceive on the face of the washer. This is the dominant element of the product interface. The next circular object you perceive is the control dial with a red outline (which selects the duty cycle: whites versus colored clothes), the subdominant object. The last level of perception is called the subordinate entity, and in this case that's the smallest grey circle, the start button. Your order of perception is guided by the size and shape of the controls. Note that this also mirrors the task you wish to accomplish with this machine—the "how," or the way you wash clothes. The first thing you'll want to do after carrying in a heavy basket is to unload the basket and load the machine. The next thing will be selecting the wash cycle with the control dial, and finally, after selecting the cycle, the last thing

you'll want to do is start it up. A good design is intuitive and delights when it is engaged. It's operation is so intuitive that it's often labeled as "common sense."

What about another example. Let's take the interior of this car (a Range Rover Velar 2019). You could carry out the same circle analysis (largest, medium, and smallest size) and see how that correlates to the function of the vehicle. Look at each circular control, from the steering wheel, to the instrument dials, to the air-conditioning and center console knob, then compare that to the activities they command.

What about some other applications of the art elements? How is color used in this interior? Do you see how the designers have employed color and shape differences? Take the strong bands of white and black, which can be considered horizontal, colored lines. These lines further separate the upper and lower parts of the dash.

The dark parts reduce glare. What about the use of silver lines and circles (rings) to outline certain shapes in the interior to accent them? Note how the stitching and seams along the dash and steering wheel create very subtle lines in the interior, helping to lead the eye and subtly separate spaces. Can you see how the different textures—chrome, leather, hard plastics, glassy displays—also provide distinction to the parts of the vehicle and cues for what controls to touch and when?

When a designer applies artistic elements effectively to the aesthetics, product detailing, and overall styling of a vehicle interior to communicate a specific feeling (in this case, sophistication and elegance) and to facilitate intuitive use, they are practicing the habits that are the hallmark of a good design.

OPPOSITE: Washer–dryer machine.
ABOVE: Range Rover Velar interior.

TOOL: THE ELEMENTS OF ART | 95

Now it's your turn! Download the template opposite and try to design your own display for an upcoming weather report. You have the artistic elements to help you organize the way the information looks, as well as the way the information is displayed. Just as a thought starter, some typical weather functions and interfaces are provided below.

Potential functions
- Current temperature
- Precipitation levels
- Weekly forecast
- Hourly forecast
- Humidity levels
- Wind speed and direction
- Air quality index (AQI)
- UV index
- Sunrise and sunset times
- "Feels like" temperature
- Alerts and warnings
- Radar maps
- Visibility
- Dew point
- Historical weather data
- Wind-chill and heat index
- Pollen count
- Moon phases
- Interactive weather maps
- Bad-weather notifications
- Holidays or local events
- Natural phenomena
- Famous landmarks
- Hobbies (surf/fishing report)

Sample interfaces

Download the template and then post your interface by using the QR codes.

96 | CHAPTER 2: DESIGN BEHAVIORS

"Technology gives us power, but it does not and cannot tell us how to use that power. Thanks to technology, we can instantly communicate across the world, but it still doesn't help us know what to say."
English rabbi and author Jonathan Sacks

2.5 | Rapid Iteration

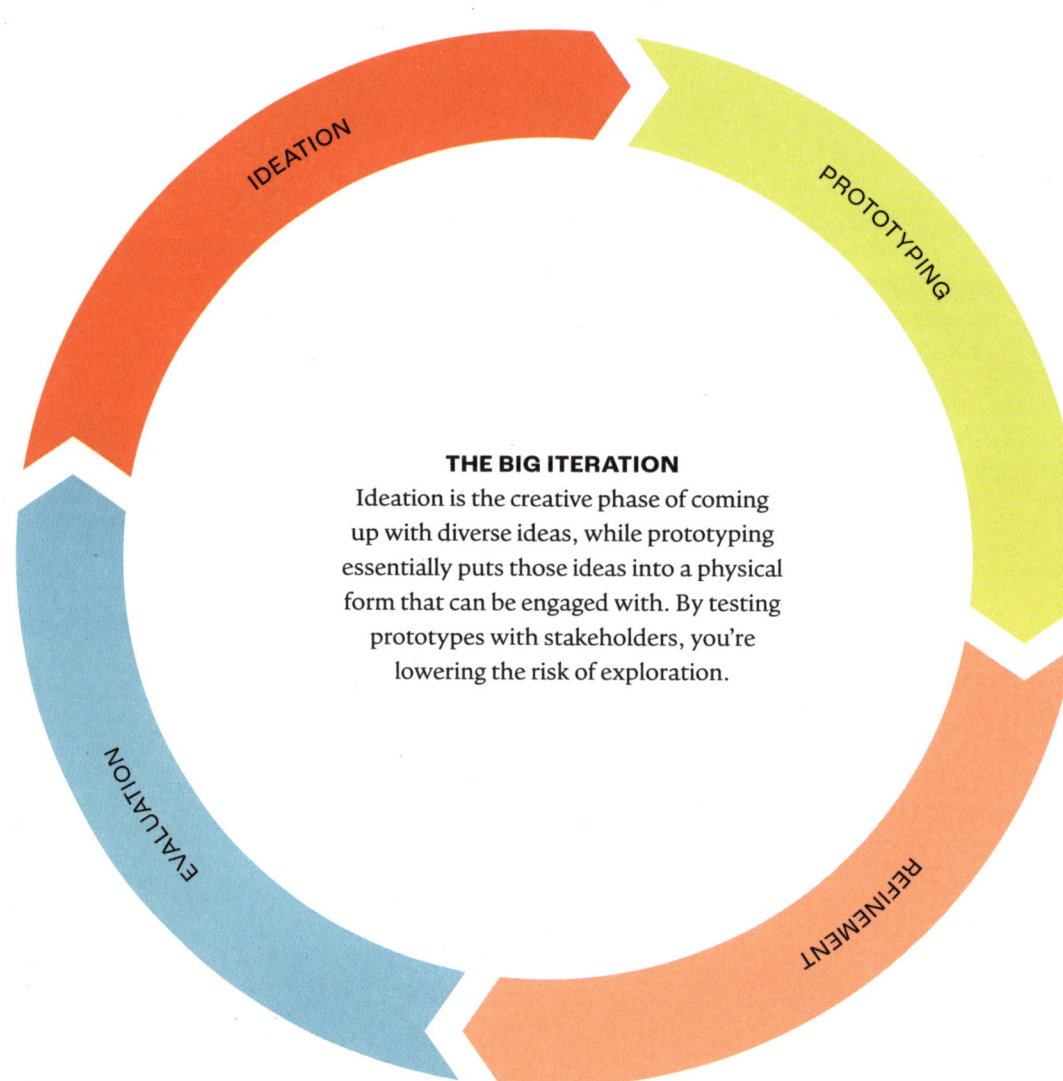

THE BIG ITERATION
Ideation is the creative phase of coming up with diverse ideas, while prototyping essentially puts those ideas into a physical form that can be engaged with. By testing prototypes with stakeholders, you're lowering the risk of exploration.

ABOVE: IDEO's Big Iteration cycle is a process for rapidly refining your work with your audience in order to arrive at a final solution that will resonate with them.

In one of my first jobs designing at IDEO, there was a commonly quoted mantra that crystallized rapid iteration as a design behavior for me: "Enlightened trial and error succeeds over the planning of the lone genius." This means that when designing, the context and use case of the design is often overlooked, especially by large organizations that are structured very heavily around rank. The assumption is that the highest-ranked person is the most expert on the business. The structure of the organization dictates that the officer with the most seniority is also the most knowledgeable about the end customer. This is actually not the case. It's usually not the highest levels of the company that have first-hand, detailed knowledge about the plight of the end customer—their pain points and negative experiences.

If you're the CEO of an airline company, you have at your disposal a private airfield and a private company jet ready to go whenever you wish. Even if you were to fly on one of your company's commercial aircraft, you would be chauffeured directly to the airstrip (no battle with parking), and because everyone in the company would know your face, you'd be brought around airport security directly to the tarmac and seated in first class. Travel for you and your family would be effortless. That's not the usual experience of the harried family with three children and multiple checked bags,

BELOW: An airfield with an array of different commercial carriers awaiting passengers and cargo.

2.5 RAPID ITERATION | 99

"THE DESIGNER COMMENCES ON A JOURNEY OF DISCOVERY, VALIDATING THEIR OBSERVATIONS AND ABSTRACTIONS ABOUT THE CUSTOMER … WHILE ALSO JETTISONING FALSE ASSUMPTIONS AND INACCURATE, PRECONCEIVED NOTIONS."

jostling through numerous checkpoints and security lines. Simply put, customer issues are often assumed away by people who are "supposed" to be experts on the subject based on their rank in a company.

There is another reason for practicing rapid iteration, in addition to misplaced faith in a company's higher management: premature misunderstanding of the customer or audience for your product. Highly trained, highly educated people often have a cognitive bias—the Dunning–Kruger effect, where the highly skilled assume that the things they find easy are easy for others (and, funnily enough, that the unskilled are so incompetent that they're unaware of their own stupidity). This means that people who understand their own products very well often wrongly assume that those products are easy for others to use. The inventor of a product might have lived with their device for months on end (often in a lab/garage) before release, providing them with the experience needed to be able to operate it effortlessly. If they never engage actual customers, belief in the usability of their device will be overinflated. In customer focus groups, the phrase "But you're using it wrong" tends to signify this trait.

Rapid iteration, then, is the fast cycling of the design process, connecting with the customer through multiple engagements, and refining and repeating until arriving at a resonant solution. It involves starting from a nugget of understanding, and progressing through ideation, prototyping, and then evaluation with the customer. The new information gleaned from the use case then inspires subsequent refinement of the idea, which leads in turn to a new evaluation and a repeat of the cycle. When done quickly and with the target audience, the designer commences on a journey of discovery, validating their observations and abstractions about the customer and context of use, while also jettisoning false assumptions, and inaccurate, preconceived notions. Personal biases and prejudices don't withstand the light of actual use by participants either, so directions that may seem plausible in a cloud of ignorance are quickly exposed to be ideas built upon a shaky foundation.

Critical to the correct practice of this behavior is to start with a beginner's mind. Try to put all preconceived notions and previous experiences away so that you can look at the situation anew. In this way, any assumptions of what you may think you know will be exposed as just that. Further, from an empathetic point of view, you'll also get to experience the joy of discovery and learning through the evaluation that the customer provides, and you'll be able to key into those insights to inspire new experiences. Work through iterations by consulting subject-matter experts, practicing interviewing techniques (covered in Chapter 3, pp. 148–55), and exploring the social sciences for practices on community engagement.

2.6 | Entrepreneurial Sustainability

ABOVE: Entrepreneurial sustainability understands creative development through its myriad stages, from concept through to mass production.

Businesses have a lifespan and they run along a trajectory. Entrepreneurial sustainability is a behavior that successful creative practitioners use when creating new services, products, and experiences. It entails striking a balance between resources and market demand.

When a business is still in its infancy (just an early idea, perhaps a technology developed with a small cohort of people), resources are limited, and name recognition of the product is nonexistent. Cash is scarce, but long nights and "sweat equity" are plentiful. Making smart decisions on your capital (financial and human) will be critical to bringing your business to profitability. This means prioritizing a customer base that's willing to adopt and pay for a novel, unknown technology or new experience. Entrepreneurial sustainability matches that customer base with a product built at the right production volume to address very specific needs. As a necessity, this customer pool will be a very small segment of the market, comprised of your most ardent and enthusiastic partners. They'll be the ones to tell their friends about your product and help bring more customers to the table. They'll also be willing to troubleshoot problems with you as you test new features or prove your product. They are your surrogate marketing team.

As your product takes hold of the market and the good word reaches more people, your customer pool will expand, and expectations for your product will change. Once you're catering to greater numbers, customers will no longer be willing to be your "technology test subjects." This larger cohort will have other expectations, such as reliability. They are buying based on the word of their friends and your reputation as a leader of products in a competitive market.

Eventually, as your company enters maturity, you'll be targeting the few remaining parts of the market—those who are more conservative and price-conscious. It's at this end of the market that products become seen as commodities rather than innovations; as a company, you are now competing on price or diversifying your offerings for every last customer. This change of customer segment is based on Everett Rogers's Diffusion of Innovations (see p. 104), a social-science theory on the adoption of new technology (in products). It describes each type of customer segment that buys into a technology, their usual proportion of the market, and the point at which they are most likely to adopt.

When you first develop your product, you are solving an issue for peers who are probably well known to you—"innovators," technology enthusiasts who share a fascination in what you're doing. You may all be united in the belief that this novel technology will help change society, but you probably don't have the disposable income to afford the product yourselves, nor its mass production. You are the "explorers" investigating the promise of the technology's implications, and targeting an extreme or specific case within the market.

The "early adopters" are the visionaries who see the promise of your idea and how it will affect society. They are the ones who will evangelize your product. Early adopters are willing to buy immediately and they'll suffer the growing pains of a company's nonexistent service, and the rough edges of a product that's not yet fully polished, because they believe in it. They expect that being the first to buy will allow them to exploit the new technology, gaining a competitive advantage in the marketplace and a foothold on which to grow. They will

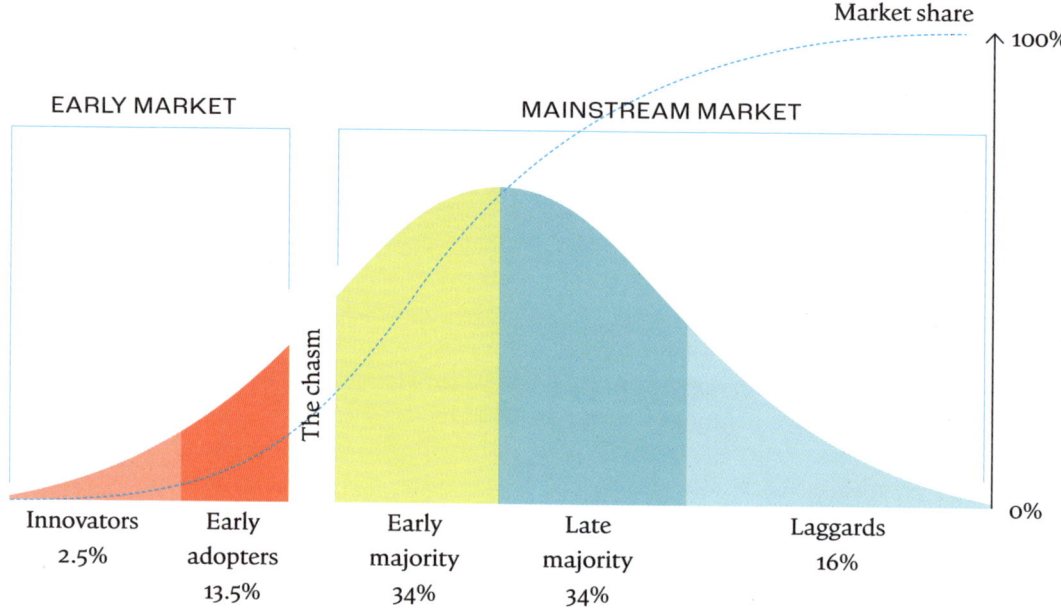

ABOVE: Everett Rogers' Diffusion of Innovations—also known as the technology-adoption curve. Many startups fail to understand the lessons that shifting customer mindsets teach.

usually demand customized features and/or modifications of your product that will benefit the next set of customers: the "early majority."

Entrepreneurial sustainability takes advantage of this concept, scaling a company's ability to adapt to the size and nature of its customer base, but the chasm between the early adopters and the early majority is a difficult one to cross. Many startups fail because they're not responsive to a changing customer base and the expectations of a larger, more mainstream market. Products and services must now take into account the fact that the early majority want to believe they're buying the best, and that they prefer to buy from a market leader who is using the dominant (preferred) technology. They believe the word of their peers (who have hopefully been influenced by the previous evangelists).

The "late majority" (or the conservatives) and the "laggards" (or skeptics) are the last groups to adopt. They are the most price-sensitive and demanding of the product. Because they're late adopters, they assume that the technology has proliferated throughout society, and that pretty much every company has the same technology within their products. They assume that the service or experience for every product of its kind will be identical, so they're now focused on price, wanting the best deal.

As a company, you can see how you would scale a product mix based on this concept. Startups develop in stages; they use the work of a previous iteration to justify the finances to mature further. In this way, they obtain more finances to make more diverse offerings, at larger volumes, to address a larger portion of the total market. At each of these financial stages, founders will sacrifice ownership for larger and larger sums of capital. To maximize ownership retention, a sustainable business will appropriate resources to create just enough

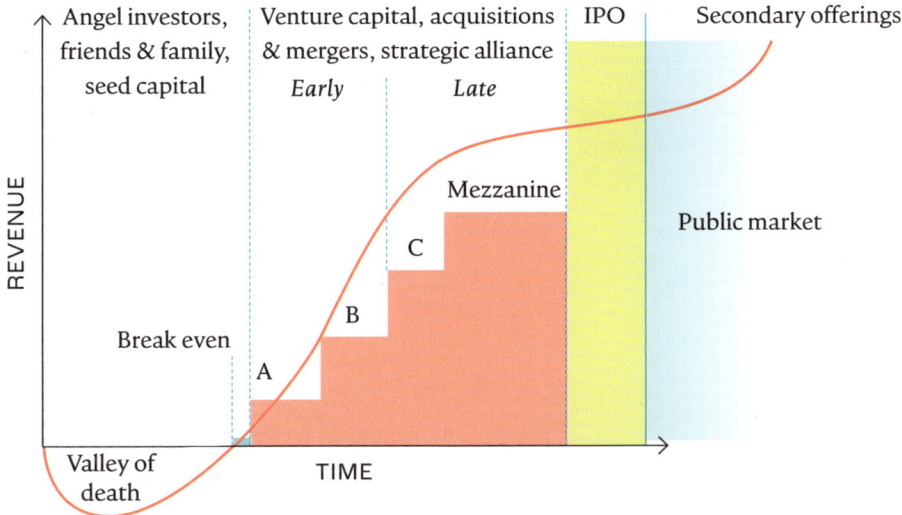

ABOVE: The Startup Financing Cycle. Startups are staged and experience rounds of investment, achieved through equity coupled with deliverables met.

products that cater to specific customers' needs based on the size of the market, and it will use data from the evaluation of that customer base to justify expansion.

A corollary to technology diffusion is the Startup Financing Cycle (shown above). Startups are private entities that initially have very few customers but plan to use their design process to prove their product with enthusiasts (extreme users) and convince the early adopters of the promise of their work. They'll seek out financing from angel investors, friends, and family to create their first minimally viable prototype, which will be a single stand-alone offering made in small batches catering to tech enthusiasts and visionaries. If executed well, the results from that initial batch will achieve the next level of financing, which will help them expand their market reach further. However, as a customer pool expands, and more revenue comes in, a company must expand (through human capital, research and development capability, and manufacturing bases) and it must also change the way it addresses its audience. Moving from a small startup with minimal cash, to an early-stage acquisition target, to a company on the verge of an initial public offering (IPO) mirrors the same technology-diffusion transition of catering to a few wealthy, enthusiastic early adopters, to a larger, early mainstream customer segment as the company grows into a public entity.

Entrepeneurial sustainability involves conditioning your building practice to match your target-market penetration (volume of buyers). By applying this behavior, you can apply the amount of resources needed to yield just the right amount of product to meet market demand, and establish a resource-rich position to scale your product mix as you grow.

Tool | Hierarchy and Principles of Design

Having established the elements of art earlier (see pp. 92–93), let's look at a few more ways in which these tools can be used to create a simple composition.

Let's start with a line, but instead of a solid line, this will be a line comprised of tiny circles:

Now let's build upon this by creating more lines:

What if we were to add some visual interest to this composition? We could do this in myriad ways. Let's take these compositions of dots and play around with how they are presented:

Space

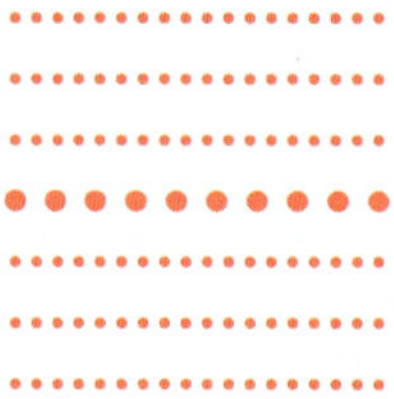

Scale

Alignment

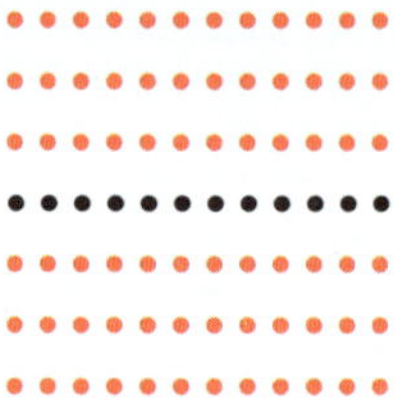

Color

Orientation

Grouping

Reversed

Quantity

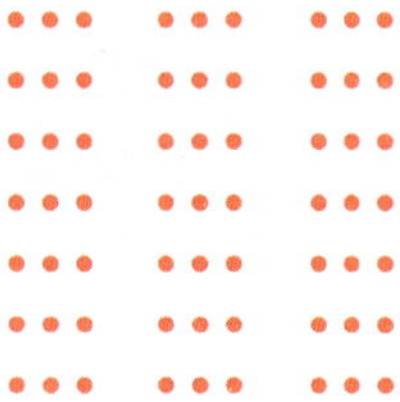

Repetition

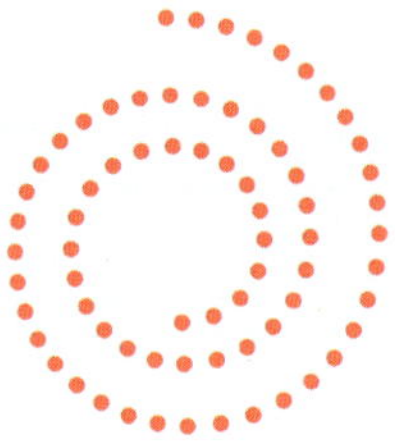

Spiral

Bleed

Variety

106 | CHAPTER 2: DESIGN BEHAVIORS

Visual hierarchy

Okay, so what's the point of exploring this degree of variation in the presentation of dots? Let's put a practical spin on the dot compositions by assigning text instead. What do we notice about the information now? Note how, when we introduce more variation in the composition, it either emphasizes or de-emphasizes some of the text. When we vary the elements in a design, we can alter the importance of them, which in turn can affect the viewer's access to the information. This concept is called visual hierarchy, and it's grounded in Gestalt theory (see pp. 110–11) and the principles of design. Gestalt theory proposes that people see visuals in a particular order: first color, then size, then alignment, and lastly language.

Principles of design

Visual hierarchy is the order in which the human eye perceives what it sees. Designers therefore employ visual hierarchy to guide the eye to information in a specific order for a specific purpose. This is accomplished through an adept use of the principles of design, creating an order using the visual contrast between forms in a field of perception, as outlined below.

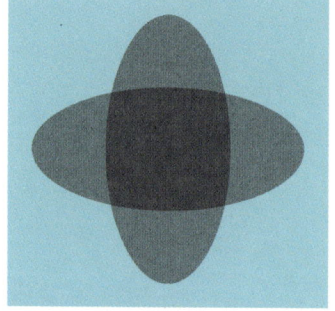

RHYTHM

Rhythm is the repetition or alternation of elements, often with defined intervals between them. It can create a sense of movement, and establish pattern and texture. There are many different kinds of rhythm, often defined by the feeling evoked when looking at them.

A regular rhythm occurs when the intervals between elements—and often the elements themselves—are similar in shape or size.

A flowing rhythm gives a sense of movement, and is often more organic in nature.

A progressive rhythm is created by using a sequence of forms that progress through a series of steps.

BALANCE

Balance is the arrangement of objects in a given design as it relates to their visual weight within the composition. It usually comes in two forms: symmetrical and asymmetrical.

Symmetrical balance occurs when the weight of a composition is evenly distributed around a central vertical or horizontal axis, or radially around a central point.

Asymmetrical balance occurs when the weight of a composition is not evenly distributed around a central point or axis.

PROPORTION

Proportion is the comparison of dimensions or distribution of forms. It's the relationship in scale between one element and another, or between a whole object and one of its parts. Differing proportions within a composition can relate to different kinds of balance or symmetry, and can help establish visual weight and depth. In the example above, note how the smallest element seems to recede into the background while the largest element seems to come forward.

EMPHASIS

Emphasis determines a composition's visual weight, establishes space and perspective, and often determines where the eye goes first. Emphasis is achieved using the various elements and principles: shape, line, rhythm, and so on. Focus or depth of field pushes/pulls your eye and therefore your attention.

Note where your eye travels within this image: first to the figure (the circle), then to the square, and last, the rectangle. Now ask: What visual principles or art elements are at work here to guide my perception in this way (contrasts in value, shape, space, and so on)?

CONTRAST

Contrast is a design principle that uses different or opposing visual elements to create focal points, add variety, and highlight important information.

Contrast can be created using many different elements:

- Color: warm vs. cool
- Value: light vs. dark
- Proportion: big vs. small
- Shape: geometric vs. organic
- Texture: smooth vs. rough
- Type: bold vs. italic
- Alignment: ordered vs. irregular

HARMONY

Harmony describes the relationship between the individual parts of a composition and its whole. It applies to the aspects of a design that are used to tie the composition together and give it a sense of wholeness, or to break it apart and create a sense of variety.

Gestalt theory

The school of psychology known as Gestalt theory was founded by Max Wertheimer, Kurt Koffka, and Wolfgang Köhler in the early 20th century. It was a reaction against structuralism, a psychological viewpoint prevalent at the time, which proposed that the mind's thoughts and perceptions are all composed of smaller, lower-level sensations or components. Much like an atom, the structuralists posited, all complex ideas could be built from smaller ideas.

The Gestaltists took an alternate view, naming their own system with a word that is used in German to refer to how something has taken "form." They believed that the brain has an innate tendency to structure "individual elements, shapes or forms into a coherent, organized whole." In addition, we perceive a structured whole ahead of its constituent parts. As Koffka put it, "The whole is something other than the sum of the parts." This "perceptual grouping" is governed by the principles outlined below: proximity, similarity, continuation, closure, prägnanz, and figure/ground.

To see how this works in practice, gather a bunch of pennies and toothpicks. We'll use these objects not for their function, but as a deliberately limited palette for exploring the principles of design. Arrange these objects on a piece of paper to create a composition. Your goal is to choose and convey one of the following topics: 1) anger, 2) ceremony, or 3) festivity.

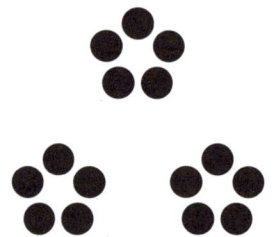

PROXIMITY
Objects clustered close together tend to be perceived as belonging together (related). These dots are therefore perceived as 3 groups rather than 15 individual dots.

SIMILARITY
Objects that are similar tend to be perceived as related to each other. Here the similarity of the circles and squares helps us define four quadrants within the square.

CONTINUATION
Objects that suggest movement are perceived as being related or connected to each other. This progression of dots suggests a line or path of movement.

Download the template and then post your composition by using the QR codes.

CLOSURE

Objects suggesting shapes are viewed as closed or complete, regardless of whether actual lines are present. We perceive a triangle in this negative space, enclosing the three sides in our mind.

PRÄGNANZ

Translated as "good form," this defines perception as reduced to the simplest forms or shapes possible. While we could perceive a square and circle here, we tend to see a single black outline.

FIGURE/GROUND

Images break down into either the figure (the positive shape: goblet) or the background or environment they inhabit, known as the ground (the negative shape: two faces).

TOOL: HIERARCHY AND PRINCIPLES OF DESIGN | 111

3
EMPATHY
Cognitive and Emotional Empathy, and Compassionate Concern

AN EMPATHETIC MINDSET IS CRITICAL TO DESIGN. IT CAN BOTH INSPIRE AND FOCUS YOUR DIRECTION. WITHOUT IT, YOU CANNOT CREATE.

3.1 | An Interview with David Kelley
3.2 | Needs versus Solutions: Reframing around People
　　　Tool: Empathy Map
　　　Tool: Maslow's Hierarchy of Needs
3.3 | Cognitive Empathy and Positionality
　　　Tool: Mind Maps and Social-Identity Maps
3.4 | Emotional Empathy
　　　Tool: Method Acting for Design
3.5 | Compassionate Concern
　　　Tool: Analogous Research
3.6 | Narratives and Interviewing
　　　Tool: The Open-Ended, Semi-Structured Interview
　　　Tool: Journey Mapping

3.1 | An Interview with David Kelley

A preface for empathy

David Kelley is the Donald W. Whittier Professor in Mechanical Engineering at Stanford University. Along with Bill Moggridge, he cofounded IDEO, one of the most influential design firms in the world. He also founded the Hasso Plattner Institute of Design (more commonly known as the d.school) at Stanford. His work originated the project-based design methodology known as design thinking, or human-centered design, which he employs both as a design consultant and as an educator. Design thinking is based on building empathy for user needs, developing products through iterative prototyping, and futuring ideas through storytelling.

When I first met David in the mid-1990s, I was a young employee of IDEO Product Development in Palo Alto, CA, and this was when I was introduced to IDEO's philosophy of user- or human-centered design. The principle of putting the human condition at the forefront of the design process greatly influenced my own practice, along with the tenets of design thinking expressed in the triple Venn diagram for innovation: a balance of humanity, technology, and business (see pp. 64–65).

After a five-year stint in automotive design, helping to design futuristic concept cars as well as production vehicles for Ford Motor Company, I was reacquainted with David in graduate school, as the founding graduate class to help establish design thinking at Stanford. He has influenced me for almost three decades, as my first design boss, graduate school advisor, and mentor. I was very fortunate to be able to sit down and interview him for this book, focusing on design thinking and the role of empathy within design.

Li: In your book *Creative Confidence* [2013], you talked about large organizations that find it difficult to be creative or empathetic. Why do you think this is? In your experiences with IDEO, how have you managed to be successful in changing this kind of organizational behavior?

Kelley: It's understandable that they're risk-averse. They have good reason to be because you can bring the company down by going in the wrong direction in a lot of places. But I really think you still need innovation, and it's just not set up that way. I mean, a hundred times in my IDEO career, I'm in a meeting with the CEO and they have the person in the room who's got the big, big moneymaking product going in the company, right? There's the guy I'm working for, who wants to do something completely new. Well, the guy who's got the moneymaking thing says, "Look, just give me a few more million dollars and I can get going making more revenue." It's just that that's shortsighted because that will end. It's hard for the company when they just got a sharp pencil with respect to money. I'm sorry, but they're basically barbaric about shareholder value and not so much about innovation as a group. Some are, but the dilemma becomes one where you want to allocate your resources between the stuff that's working now and the new stuff. So how do you get that going?

IDEO, as the group trying to do innovation, tries to get a few small successes, pick something small. This is my religion so I'm overly positive about it. We tend to be able to go in and find something that's a quick hit. It's small and surprises people because you know, it fits. The Venn diagram is a good idea because it's got a technology and it's a business idea. And it takes people into account. So if you can, get a small win. When you try to get people to do a big project, it's just really hard. But, if you get a small project and have a small success, everybody looks at that and says, "Look at that, I didn't see that coming." And then you get a little more credibility and you build a new building.

Li: Speaking of small successes, you mentioned something in your book about thinking in a more human-centered way. You talked about how Doug Dietz was the lead designer at GE Healthcare. He developed an MRI machine that they loved, but at first it just didn't work well for customers. Then, after learning about this human-centered design process, he was able to dramatically improve the take rate among children—80 percent of child patients had had to be sedated in order to be scanned, and afterwards they didn't have to be. Tell me a little bit more about that.

Kelley: Okay. So first, when Doug and his team designed the machine, they had no idea about Stanford or anything to do with design thinking—they just did it the normal corporate way. And it was a technological marvel, how many lives it saved; it was doing a diagnostic test that you couldn't get any other way. And it was finding these cancers ... So they hung their hat on how wonderful the technology was. That's a corporate thing to do. That's a reasonable, understandable way to do it.

And then, see, here's my line: Basically, people are going to have to weave this into the fabric of their lives in order for it to be successful. All the big successes you can think of—the internet, the iPhone, AI—will work when they're woven into the fabric of people's lives. So you can have this spectacular medical technology, but if you don't make it so that it works in people's lives, both physically and emotionally, then it's not going to be successful. It's certainly not going to be the breakthrough that the company wants, given how much they've invested in the thing. It's a great example because they got the technology and the business idea right—they could charge a fortune for it. But, they just didn't have the interface right, especially for kids.

So once Doug came to the d.school and got the notion of human-centeredness, it just resonated with him. I mean, everybody was looking at the numbers of how successful the thing was. They weren't hanging around the machine. They'd already delivered the machine to the hospital. It's not their job to hang around. So after the d.school thing, user interviews and user observation—mostly observation in this case—he went out and looked at it and found this problem with kids having to be sedated. And once you have that—once you have the need, a real need—well, it turns into problem-solving and we're actually really good at that. We're just not good at need-finding. So once you see the issue with kids, then there's a creative leap that has to happen, which is why designers were involved. They made the creative leap that we should make it enjoyable for kids. It was torturing the kids at that point; the opposite of that is making it enjoyable. So you build what they call the Adventure Series. It's like going to Disneyland and going on a ride. It's not that different from going on a ride, you know. You get in, it drives your body in and out of the machine and makes a lot of noise and all that stuff. So it's a really good example of having the other two circles and missing the human one.

Li: You and IDEO's designers want to do things in first person, right? You mentioned walking in the shoes of others [which we call moccasins; see p. 142]—washing other people's clothes by hand, or staying as guests in housing projects, standing next to surgeons in the operating room, in order to design around their lives. It gives you a personal connection with the people you're designing for. Why is this first-person connection to humans so critical in business and in design?

Kelley: It goes back to what I said before. For this to be a breakthrough idea, whoever your user is going to be, it has to fit into the fabric of their life. They just can't like it—it has to really fit. So you have to see where they eat

"PEOPLE ARE GOING TO HAVE TO WEAVE THIS INTO THE FABRIC OF THEIR LIVES IN ORDER FOR IT TO BE SUCCESSFUL."

and what they do when they're not working—all that kind of stuff. You just learn things.

You were talking about standing next to surgeons. IDEO was tasked with improving an indeflator for angioplasty. There's a little tube that blows up, and there's a machine connected to the other end of the tube, which you use to pump the balloon up. All the existing models were one-handed. The doctor had to do it all with one hand, which made it really difficult to set the right pressure and all that stuff—they were doing it, but it was cumbersome. Anyway, we were charged to do the next version, which everybody assumed would be one-handed because it seemed that was a hard, fast rule. Anyway, we go and we stand there and watch them use it, and they're not using their other hand for anything else. So we made one where you hold it like this [two-handed] and then you can systematically use the dial and tune it right in because you don't have anything else going on with the other hand. I don't think you'd learn that by looking at the spec of the old one and trying to improve it. You basically take that as a given because the myth is that it has to be one-handed.

Li: I have definitely run into this groupthink issue myself in corporate design. Someone made a decision from ten years ago based on some situation—maybe the first person they ever thought about using one said, "One-handed might be nice." And from there it's gospel and they never bother validating the opinion. Doctors five years ago are different than doctors today. It's so important to refresh your personal understanding about the fabric of people's lives. So those types of things happen unless you've actually had them tested in the field.

Kelley: Once you're there and you've tested in the field and you understand what the needs of the physician are, in that particular case, you have the conviction to act, right? The reason nobody changes the one-handed one is they didn't have the conviction that there was any reason to change it. Being there builds that intuition and that knowledge that compels you to act. I'm sure they sat around and played with the indeflator, but they weren't in a situation that was real enough that they'd realize they could use the other hand.

Li: So it's like you said, getting that deep understanding of people's lives in first person helps you method act their life, help understand what they go through, more so than just sitting at your own desk or cubicle, guesstimating what you think they go through. You actually do it.

I'd like to follow up with the role of the designer. Which is to help translate that human requirement as well as, as you said, to perform the creative leap, right? That's why you have a designer there as well. In codesign, does the designer help facilitate and translate those needs into the physical or the viable or the visual? Codesign is similar to human-centered design. The role of the designer is to be the person who helps actually translate something into a design. So if you think about a solider having a human requirement for a refrigerator, for example—the soldier doesn't know how to create a refrigerator, so it's not his job to draw and model and machine it, but you're taking on his requirements, what he wants out of that appliance.

Kelley: They have the mindset that they conceived of the need for a refrigerator without having the responsibility of designing it. I think it's in the acceptance of it that it makes a real difference. So that ownership of the idea is really essential in adoption.

Li: Yeah. It's about trying to find a way to build equitable consensus. I think that's one of the beauties of what human-centered design does—especially in social spaces where there aren't the same equity levels and you're really trying to design with these different levels of empowerment in mind. You're trying to empower those who don't have power. So in a lot of ways, human-centered design, to me, is trying to better society. That's what we do, right? By trying to represent and help encourage the voices of those who may not have a voice at that table. I love that idea.

Kelley: And it's not just developing-world stuff. We went to do a project with [healthcare provider] Kaiser where, when we interviewed everybody, it was pretty clear that the nurses were the key to better healthcare, but they didn't really have a seat at the table with the administration or the doctors—but they did in the end. If you just go through and chronicle all the things that have to happen—who's the one who has the understanding of the patient?—you can make a human-centered design product or service. Once we shone the light on the nurses then they had more agency. They were then anointed in Kaiser, which is still true today. They were kind of the innovation group, but the nurses were where the innovation was going to come from. That was accepted by the higher-ups and it's worked really well.

Li: Thinking back a little bit now, you've had a very long, storied career. You created IDEO in the early 1990s, educational programs like the d.school in the 2000s, and have engaged in the development of countless companies, startups, and initiatives. How has your design approach changed over time? Has it changed based on the entity or the industry? Has it changed due to the time or the zeitgeist? The culture? Was the process always the same, or were there some differences between ecosystems and industries?

Kelley: Well, I hope we've continuously improved the methodology, the process that we use, but it's not the value system. You articulated the value system when we first started talking, when you rightfully talked about what our values are. So the value system has hardly changed at all. Stuff like human-centered design and the value of prototyping and storytelling, that's been there since when you were there in 2005. It's still there as far as our values. But the details of how we work change a lot; the tools and the opportunities change. And that drives you to a different methodology—not different values, but a different methodology.

So now we're in the middle of the AI revolution, and that's driving everything that we look at—what's the mash-up between AI and our different clients or our different industries? That's completely different than trying to figure out how to make a pool cleaner, back when we were doing that. So yeah, it's changed; it changes the complexity. But I think the statement I would make that's the most unusual or interesting is that our values haven't changed.

Li: I would agree with that too. I think the ethos we believe in, as far as our philosophy on design goes, will always entail these ideas about understanding people, how they live, and trying to design something that makes things more pleasurable, more equitable, more fun. But the ways in which we do those things will change. When you think about AI utilizing generative design in which you can now crowdsource a thousand designs—how do you maintain your sense of what your design should be?

Kelley: Yeah, well, AI is a tool. Every time I want to write something or come up with ideas, I go to ChatGPT and it makes my ideas better. It comes up with ideas I hadn't thought of. But back to the creative leap. There's still a career in design: Which one of these is going to work? Which one's likely to enter the fabric of people's lives, and which one is going to get across to them? Going to the triple Venn diagram, which one's going to be at a price point that'll work? Or, which one can we find a way to distribute from a business point of view and get it to the people? And, which one has a technology that's robust? You've got all those things that require a creative leap, even though you have the data.

Li: Yeah, you have the data of all the designers. You have the bell curve of their ideas. And the interesting thing is that sometimes it's not right that you would go with the median. Sometimes it's the extremes that you're looking at, right? It's still the designer's role to look through all that data and ask, "What fits best?"

Kelley: Fit isn't something they could figure out. But, one thing that AI has really helped me with is mind mapping. I'm a huge believer in mind mapping, so my point of view about my mapping is that brainstorming is incredibly powerful, especially if you mean brainstorming in the way that we mean it. Presently, "Let's have a brainstorm" means "Let's have a meeting."

Li: [Laughs] That's right! Yeah, it's been corporate-filtered.

Kelley: The quintessential part that we care about with brainstorming is building on each other's ideas. Anyway, a mind map is the equivalent of a brainstorm from an individual point of view. ChatGPT is really good at that because what you're doing is trying to push further away from the norm. Every time you get a little way out, you can then use ChatGPT to go further because it's got more access to possibilities. So mindmaps are getting better.

Li: The last question I have comes from one of our beloved mentors, Matt Khan [pioneer in design coursework at Stanford], who used to say, "How have you used design to design?" Looking back, how do you use design to design your life, or just use design to design?

Kelley: It's like my religion. I'm asked to use design to do everything. I'll give you one example, which is when my family comes together for Christmas Eve at my house. We do the normal things—eat and sit around and talk—but, I thought, it could be more fun, we could get more engagement. So a few years ago I started designing something that I called the Kelley Olympics. I bought these stupid little plastic gold medals, and instead of just sitting around we'd go around and see who could stack the highest stack of Oreo cookies. If you did that, you got a medal. Or, who could spit a cherry pit the farthest. There were other things too. And it was successful. It was designing the experience, you know, designing the Christmas Eve emotional thing.

Li: Yeah, I completely understand that as far as its resonance goes. What I've always thought is, design is not a discipline; design is a lifestyle. You let it permeate everything you do. I'm similar. I designed my house, I decorated my interior, designed my dining-room table, and I designed my wife's engagement ring. Everything that I can touch, where I know what I might do or make may improve the experience or the life of someone I care about, I'm not going to want to just buy something off the shelf and give it to them— I'm going to want to design it.

Kelley: I'm surprised how far out of the norm we still are, right? I designed my wife's engagement ring, but name somebody else who's done that, other than you, other than a designer, you know what I mean? Everybody goes to the place and buys the ring.

Li: Yeah. I think there's a little bit about creative ownership. And that level of detail— I know my wife the best and so I want her to be the happiest she can be and I'm not going to entrust this to someone else. I studied jewelry design at Stanford, but I've also met people from Tiffany and I know who their suppliers are. So, I mean, some people may not want to go through that triple Venn

"HUMAN-CENTERED DESIGN IS ... TRYING TO REPRESENT AND HELP ENCOURAGE THE VOICES OF THOSE WHO MAY NOT HAVE A VOICE AT THE TABLE."

diagram. Some people may want that convenience of walking to a store and picking something up and trusting that it's good. But at least personally, the joy of seeing and understanding what my wife thinks about when we talk about marriage is a great way to get more and more connected to someone. So why not use an opportunity to do that?

Kelley: It's just surprising that it isn't a tool that more people use. But it's our lot in life. We don't have any choice [laughs].

From David's interview, you can see how successful designers cultivate their intuition and practice observing and involving themselves first-person with the people they design for and with. This mindset is one of empathy and perspective-taking and it's crucial for a product or service's adoption.

3.2 | Needs versus Solutions: Reframing around People

One of the first things to remember when it comes to design empathy is to put your focus on the human experience. Framing around the human allows you to take the user's perspective. One way to do this is through the act of discovering needs, both explicit and implicit, so that you can create appropriate solutions. This is called "need-finding." Needs can be thought of as any physical, psychological, or cultural requirement of an individual or group. Something that's missing.

With any creative endeavor, the more you practice, the better you get. This is also true of need-finding—the more you interact with and have experiences among a group of people, the more you become attuned to their idiosyncrasies and ways of looking at the world. In short, the more you interact with a group or tribe of people, while maintaining an open attitude, the more you learn about their culture.

As with many cultural signifiers, there are some things that are concrete, easily observable, and explicitly displayed. Think about the type of clothes people wear to support a local sports team, or how they interact with specific objects in given situations, such as turning a door knob, drawing out cash from a purse, or using their phone to pay for a transaction. These are all easily observed actions, and they all display explicit needs.

Other actions, however, can be more subtle. These behaviors may be more abstract and less easily observed, yet they still reside within the culture and are commonly understood. These implicit behaviors may go unspoken but they're nevertheless known to the community. For example, elevator etiquette. When entering an elevator, if there's already someone inside, the "acceptable" behavior is to press the button for your desired floor, then move to the farthest corner from the other occupant. It's considered rude to stand right next to an elevator rider if the rest of the cabin is available to you. No one has to announce this when riding elevators; it's simply taken for granted that you respect other people's personal space.

One of the core principles of need-finding is that needs are framed as verbs. Considering human needs as verbs allows for greater creativity when designing a solution—which is any product, service, or experience (nouns) that addresses that need. Identifying the need represented in the picture opposite, a verb could be employed in a phrase such as, "She needs to reach something on a high shelf." This need is pretty explicit and can be easily observed. So why do we phrase it this way?

A natural tendency may just be to prescribe a solution. "She needs a ladder." However, can you see that this type of phrasing, built around a noun—"ladder"—is a deductive thought process? It's the most obvious solution to the problem and therefore inherently limits creativity. If we want to channel more creative thought processes, we might instead ask, "What are all the different ways in which we could reach something on a high shelf?" Now, we'll be freed up to imagine many different things: a trampoline, stilts, rocket jetpack, revolving shelves, asking a tall friend for help, a zero-gravity generator, revamp the store layout, and on and on. By contrast, if tasked to design a ladder, how many different ladders can you truly

> "IF WE ONLY FOCUSED ON JOBS TO BE DONE, WE'D ONLY CONTINUE TO ADD MORE FUNCTIONALITY ... AND FAIL TO REALLY UNDERSTAND THE CUSTOMER'S EXPERIENCE."

create? Sure, the ladder may be cheaper, taller, or a different color, but it's still a ladder; your creative output has been stifled.

Clayton Christensen, a distinguished expert in innovation, has coined the concept of "jobs to be done"—or framing an activity or job that an object performs. While I would agree with Christiansen's framework, this is a similar structure to need-finding in that the ladder performs "a job to be done," which is "to reach something on a high shelf or to provide reach for someone." This is similar to a function tree in engineering practice, where the job to be done by a hammer is "to drive in a nail." In other words, this framing places the focus on the object and may sometimes limit our ability to empathize with the customer.

For example, the person reaching up to a high shelf in a grocery store may be embarrassed about being short. This identifies an implicit need. Having their short stature noticed by other shoppers might lead to feelings of inadequacy or insecurity, and this could impact the outcome of the design. A tall ladder would merely draw more attention to someone trying to avoid being noticed, not to mention the fact that there would probably be liability regulations and work-safety practices prohibiting customers in a store from using a tall ladder anyway. Here the need to "reach for something in a way that doesn't draw attention" or "grab an object in the same way that taller people do" is now a descriptor for the solution, and a ladder may no longer may be the best choice. In fact, in the case of the latter need, the ladder is obsolete. If we only focused on jobs to be done, we'd continue to add more functionality to the ladder and fail to really understand the customer's experience in the store.

If you extrapolate this further, all products, services, and experiences have multiple functions, which are driven by a diverse set of human needs. Some needs drive the primary, explicit task (reaching something on a high shelf), but there are a myriad of more implicit needs driving secondary functions, such as the manner in which the task is performed (do this discreetly, or in a way that everyone can use it in the same way). In this scenario, having shelves that revolve around a track may be a better solution. Where most designers and engineers err is that they focus almost exclusively on the primary task and their conceived notion of "optimizing around that function," without bothering to understand the secondary functions that are driven by more implicit and abstract human requirements.

Can you think of a product, service, or experience where the secondary functions, driven by implicit need, were what you remembered most? A blender's primary function is to "blend ingredients into a liquid." But watching fruit get blended into a smoothie, watching the blades turn, probably isn't what you remember most about your experience with a blender. It disregards a potential secondary function of providing a party host or bartender with an opportunity to "perform an act of service" so that you can feel as though you are enjoying a tropical drink in a faraway island destination. In this setting, a bartender sees themselves not only as a maker of drinks but as a performer "displaying their craft to you" or "entertaining you through finding out your tequila preferences." Whenever you're designing and encounter a new person or situation, can you think about both the explicit (spoken) and implicit (unspoken) needs? What are the concrete (observable) things you remember about the person, and what are the abstract (subtle) characteristics that you noticed?

Tool | Empathy Map

An empathy map is a very powerful tool for understanding others. This method can be used when observing people in the field and ascertaining their needs. It can also be a tool for analysis and reflection, and for identifying opportunities for design. Let's take a look at its myriad uses.

Suppose you're designing a new piece of safety gear for skateboarding (like an extreme-sports helmet). You go to the local skate park to observe and talk to skateboarders and get a sense of how they practice their sport. You then use an empathy map to assemble your raw data: notes in the field, recording what you notice people saying and doing. Let's say you talk to a specific skateboarder—your completed empathy map might look like the one shown opposite. (Your map will have more data than my example; this is just to get you started.)

Something to note about the map is the difference between the quadrants on the left and right. Those on the left-hand side include what's explicitly stated or done, which can be recorded with a camera. The quadrants on the right-hand side, however, contain implicit data—thoughts and feelings, which tend to be hidden below the surface. You'll have to use your empathy skills to fill these in because the skateboarder won't likely tell you directly what they're thinking or feeling. You'll have to interpret their words and actions instead. Don't worry, though—you would validate the accuracy of your theories later in your design process.

After gathering raw data, it's time to take the empathy map a step further. This time, looking at the notes already in place, populate each quadrant with what you think are the relevant needs. (Again, you'll probably have more needs than I've listed here.) Note too that, as we've mentioned, needs are always in the form of verbs. The map has now become a tool for analysis and reflection.

Finally, use the empathy map to identify any contradictions and tensions that might exist between different needs. Did your subject say one thing and do another, or did they appear to be thinking something that was at odds with what they were expressing? In this example, you might pick up on the skateboarder's need to establish kinship while at the same time wanting his fellow skaters to give him special treatment. Or how about the need to play safely, but only protect select body parts? How does the particular value of "personal safety" manifest itself in their behavior? By looking for tensions or contradictions in words and actions, we're identifying frames of meaning, especially those that are either consistent with or in opposition to the rules of a subcultural group.

In addition, can you identify pains and gains in each quadrant? What does this person find difficult or uncomfortable? What are the potential ways in which they benefit from this activity? Think deeply and imagine what they might be going through. This is a critical element of empathy—getting perspective from another person's point of view. Remember that, though their behavior may seem foreign to you, or even counter to your own views, from their side, in their world, their needs make sense.

EXPLICIT | IMPLICIT

Say

Notes
"I really like the camaraderie of skateboarding."
"I always play the sport safely and I'm helping others get better."
"I'm at the top of my game—I just got a sponsorship deal."

Needs
- Establish a kinship with fellow skateboarders
- Play the sport safely
- Move up the ranks
- Obtain money or fund my passion

Think

Notes
My friends see I'm trying to perfect this trick; they won't mind if I cut through for extra reps.

I need to focus on getting my foot off the board quicker to do this trick.

Needs
- Communicate to friends the urgency and importance of who I am and what I need to do
- Focus on my body, the board, and what I need to do to improve

Do

Notes
- Leaves helmet on the bench but wears kneepads
- Cuts in front of others waiting in the queue for their turn on the obstacle

Needs
- Protect my body, or at least the parts of my body I actually injure
- Practice tricks on skate park obstacles consistently and repeatedly

Feel

Notes
I'm entitled to cut in line—after all, I've paid my dues.

Needs
- Display my experience and skill level
- Receive respect for my game
- Inspire others to achieve similar levels in their sport

ABOVE: The quadrants of an empathy map make it easier to spot contradictions or tensions between what someone might say, do, think, or feel.

Download the template and then post your empathy map by using the QR codes.

TOOL: EMPATHY MAP

Tool | Maslow's Hierarchy of Needs

This tool provides a mental framework for understanding the different categories of human needs. The American Abraham Maslow was one of the most influential psychologists of the mid-20th century, along with Sigmund Freud and B. F. Skinner. Maslow was the first to classify and provide a framework of needs relevant to modern society—presented in a paper called "A Theory of Human Motivation," published in the journal *Psychological Review* in 1943.

Maslow wrote about basic human needs falling into five different categories: physiological, safety, social (or love or belonging), esteem, and self-actualization. These were defined as essential needs for all humans, with the first four categories considered to be "deficiency needs," where a deficiency in one of them would result in a loss of well-being. Self-actualization, however, was described as a "growth need"—needed for growth but not harmful if unfulfilled. Maslow also identified needs that were of a higher order than these basic needs, which dealt with curiosity, aesthetics, and growth, but for this tool, we'll focus on the groups detailed opposite.

While most often depicted in the form of a pyramid, there's actually no such association in Maslow's writing. Instead, he suggested that the sets of needs are interconnected but that they do not cause each other; they simply coexist, with a human striving to ultimately satisfy her self-actualization needs. However, certain needs, he said, can take priority over or overshadow others, or present themselves as dominant, depending on the situation.

For example, suppose you know someone who practices Islam and who therefore fasts during the daytime during Ramadan (the ninth month of the Muslim calendar). You see that they have a physiological need to nourish their body, yet for part of every day they are sacrificing the opportunity to eat. Here a more abstract need—"honoring and connecting with your faith," which is both an esteem and a self-actualization need—supersedes the more concrete (readily observable) need to "nourish the body."

Can you think of a time in your life when you had multiple needs that played out in different ways? Some may have dominated others, while at other times, you might have sacrificed needs at one level because of another need at a different level. Have you ever been so hungry that you've become grumpy and snapped at others? This is an example of a physiological need taking priority over a social need. Or, have you sacrificed sleep (physiological) and stayed up studying in order to perform better on an exam (esteem or self-actualization)?

Download the template and then post your tool by using the QR codes.

Maslow's hierarchy of needs is a good tool for both organizing the needs you've observed when engaging with someone, and hypothesizing about more abstract needs that might lie just under the surface. Pick an individual case to focus on and then see how many needs you can identify within each category using your empathetic mindset. Though some may be hypotheses that need validation, trying to understand how some needs may work with or against others is a good habit to develop. It will allow you to develop your empathy skills and gain a deeper understanding of someone else's outlook on the world.

Maslow's hierarchy of needs

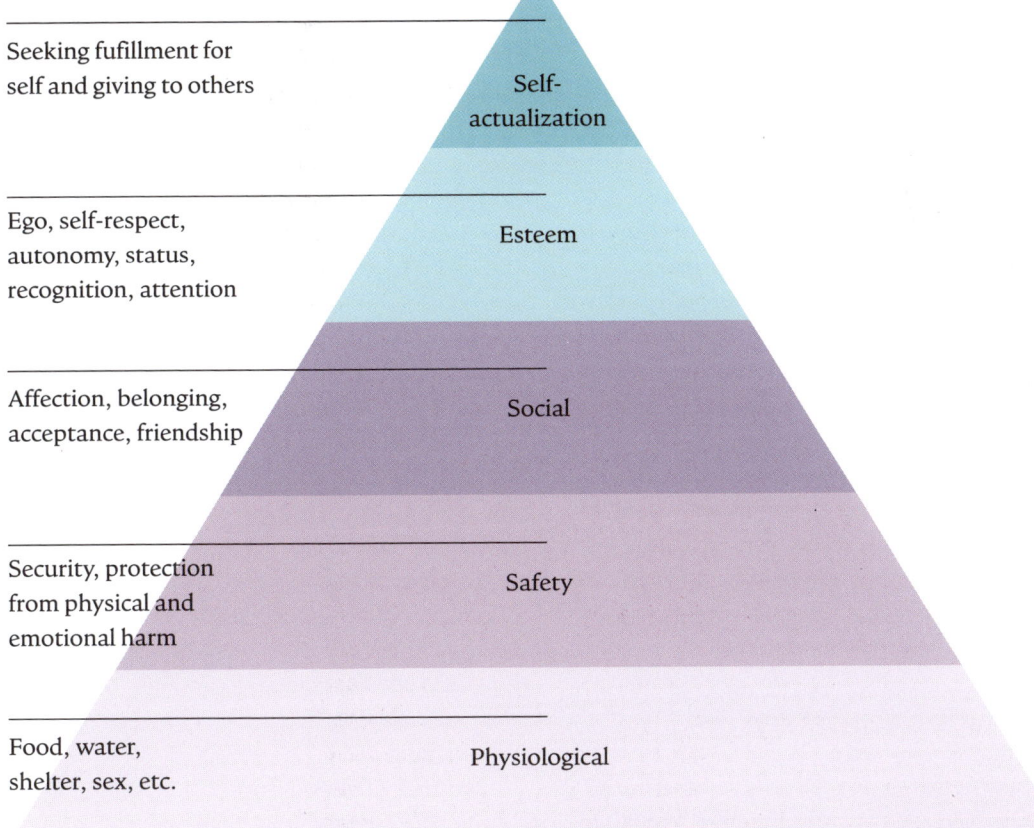

Seeking fufillment for self and giving to others — Self-actualization

Ego, self-respect, autonomy, status, recognition, attention — Esteem

Affection, belonging, acceptance, friendship — Social

Security, protection from physical and emotional harm — Safety

Food, water, shelter, sex, etc. — Physiological

3.3 | Cognitive Empathy and Positionality

Design empathy encompasses empathy in all its forms to understand an intended audience, and by using it we hope to better tailor a product or design to help improve their lives. One thing to note is that empathy is not the same thing as sympathy. Merriam-Webster defines empathy as the "action of understanding, being aware of, being sensitive to, and vicariously experiencing the feelings, thoughts, and experiences of another." Sympathy, on the other hand, means being moved by or responding to another's situation. It doesn't entail a shared recognition or an emotional experience. Brené Brown, a researcher and professor in social work, uses a hole metaphor to showcase the difference between empathy and sympathy. If you came upon someone who'd just fallen into a hole, sympathy would be shouting down to the victim, letting them know that you'd called the authorities and that help was on the way. Empathy, on the other hand, would be voluntarily throwing yourself into the hole to be with them, to understand their situation, and to experience the pain they were going through firsthand.

As mentioned earlier, Goleman and Ekman have defined three types of empathy: cognitive, emotional, and empathetic concern (see p. 84). Over the coming pages we'll consider each type in turn, and look at a few tools you can use to develop and practice them.

Cognitive empathy

Cognitive empathy is the ability to understand another's perspective. Also known as "empathic accuracy," this type of empathy refers to how well an individual can perceive and understand the emotions of another. Simply knowing how another feels or what they might be thinking (the contents of their mind) is cognitive empathy. Cognitive empathy is a skill: humans can learn to recognize and understand the emotional state of another person as a way to process emotions and behavior. Sometimes referred to as "perspective-taking," cognitive empathy is the first step toward understanding the world through the eyes of another, recognizing their point of view and how they might have arrived at their current state.

There are many ways to develop your cognitive empathy. To practice, you must first be at peace with your own outlook. If you're preoccupied, take some time to calm your mind and establish a mental state that will leave you available to perceive and engage with the emotions of another. If you want to address people's needs, you have to forget about your own problems and worry about their lives instead. Remind yourself too that it's very easy to react to people based on your own point of view. It's important to be able to hold your own prejudices and notions at bay if you want to listen and observe effectively. If you interject with a strongly held belief of your own, you'll only shut down the person in front of you. Even if it's just a matter of acknowledging that you understand them and have had a similar experience (which is a common tendency), try to refrain from interrupting because it doesn't develop your empathetic skills. I often overhear these types of conversations at social functions and I've coined a phrase for them—"resumé-boosting dinner-party conversations." For example:

> Person A: "I'm really into photography, I find the art medium a great source of expression."
> Person B: "Yes, absolutely! I'm an avid photographer too! I regularly shoot digital full-frame and love medium format 220 film! What about yourself?"

In this instance, Person B, in their rush to identify with the person across from them, 1) ignored the actual subject of the conversation, which was artistic expression, 2) presented their own experience as equal to or greater than Person A's, and 3) changed the subject to equipment and technical proficiency, of which their companion may have no experience, or be interested in discussing. Additionally, Person B may now assume that everything they know is something Person A also knows because if they're such an expert on digital and medium-format film cameras, then the other person must know all about their dabbling in custom-coded image-recognition algorithms, right? Instead of going down this path, listen, and allow the person you're talking to enough time to develop and present their own viewpoint. You'll know you're doing this well when you notice that your ratio of listening to talking has become much greater.

This is practicing a beginner's mindset. When engaging with others, try to look for cues as if you're hearing everything for the very first time. By taking a more naïve, objective view—even if you have previous experience in the field—you'll be able to observe and hear things without a preconceived notion of what they mean. When engaging with someone, focus on being present. Maintain eye contact, and observe their facial expressions, body movements, and gestures as if you've never seen them before (which will often be the case). However, even if you're talking to someone you're quite familiar with, if you remain open and attentive, you'll notice that nervous tic that you've grown accustomed to, but you'll see it in a new light. Make a note of changes in facial expression, or any other physical gestures, and how those transitions coincide with the words that are being said. Process what they're saying and follow their train of thought, but enjoy the ride—don't try to predict what they'll say next or spend time formulating a pithy response. Just let them be and give them time and space to open up to you.

In summary, listen to what people are saying, while watching what they do. Listen also to what they say about what they do, and pay more attention to what they might be thinking or feeling behind the words. Try to spend more time observing and listening than trying to interject or come up with what you'll say next. If all this is done correctly, you'll engender trust and build rapport, which is the actual point of having a genuine conversation. Then, when the conversation is over, it's important to immediately take stock of what was said, so having a small notebook with you—or a phone with a notes app—is a good idea. And drawing up a quick empathy map (see pp. 126–27) to organize what you've just experienced will pay dividends later, when you want to revisit the encounter to filter through the needs that arose.

It's worth noting that understanding and practicing all three types of empathy is important when instigating creative output. Cognitive empathy on its own is insufficient, just as emotional empathy by itself would be insufficient without cognitive. For example, if you encounter someone who's just lost their job, merely understanding their frustration and anxiety without actually internalizing their emotions will mean that you risk coming across as somewhat detached and cold. Conversely, feeling their emotions so deeply that you feel paralyzed by their situation without having a full understanding of their state of mind and surrounding circumstances is not productive either; you might start to feel so depressed about their job loss that you feel incapable of providing a solution. In this case, the balance of both cognitive and emotional empathy would be tempered by the third type, which we'll discuss later in this chapter: an empathetic, compassionate concern about the situation that motivates you to help. A designer needs to be able to channel all three types of empathy into a positive and productive solution.

Positionality

Employing cognitive empathy sounds quite straightforward—noting things that are relatively easy to deduce about someone—but there's a lot more subtlety than you might think at first glance. People's cognitive states are not formed in a vacuum. Whether you are cognizant of it or not, everyone carries some implicit biases that inform their understanding of the world, resulting from their cultural background. For example, a familiarity bias may lead you to assume that anyone you encounter who graduated from the same university as you did will be similar to other people you remember from your student years. This will not be the case. By choosing the familiar or comfortable over what is actually known by the other person, you will have indirectly inferred and mischaracterized their point of view, even if you had good intentions, such as evoking a sense of camaraderie between fellow alumni.

These biases can often derail our empathetic understanding of someone else. There's a natural tendency to assume that the influential experiences in someone's life will be identical to our own. Or, that having been raised under common circumstances through shared experiences, those influences will have played out in exactly the same way. An awareness of the concept of positionality—our position relative to others—is one way to limit the intrusion of these biases. Positionality refers to the social and political context that creates your identity, blending race, class, gender, sexuality, and ability status. Positionality also describes how your social identity influences, and potentially biases, your understanding of and outlook on the world.

Just as your mindset informs your preferred way of thinking and working, and in turn shapes your design process, positionality affects your way of seeing the world. Viewing things through your own cultural lens determines your perspective and how you frame meaning when comparing different attitudes around the same topic. For example, the role that work plays in our lives will differ depending on the country or cultural group we belong to. Do you live to work or work to live? Is a job a purposeful reason to get up in the morning, or merely something that one does in order to put food on the table? Is it polite to talk about one's vocation at a dinner party? The answers to these questions will vary, often based on positionality. We have to be cognizant of our own positionality, then, to recognize a difference in positionality in others. By seeing the differences and accepting them, we begin to see how our choices, made from our own point of view, might affect others. (On p. 136 we'll take a look at a tool that will help you develop a deeper understanding of positionality.)

Sample cognitive biases

Fundamental attribution error

We judge others on their personality or fundamental character, but we judge ourselves on the situation.
Example: "Sally is late to class; she's lazy. I'm late to class; it was a bad morning."

Halo effect

If you see a person as having a positive trait, that positive impression will spill over into their other traits. (This also works for negative traits.)
Example: "Taylor could never be mean; she's so cute!"

Self-serving bias

Our failures are situational, but our successes are our responsibility.
Example: You won that award due to hard work rather than help or luck. Meanwhile, you failed a test because you hadn't gotten enough sleep.

Moral luck

Better moral standing happens due to a positive outcome; worse moral standing happens due to a negative outcome.
Example: "X culture won X war because they were morally superior to the losers."

In-group favoritism

We favor people who are in our in-group as opposed to an out-group.
Example: Francis goes to your church, so you like Francis more than Sally.

False consensus

We believe more people agree with us than is actually the case.
Example: "Everybody thinks that!"

Bandwagon effect

Ideas, fads, and beliefs grow as more people adopt them.
Example: Sally believes fidget spinners help her children. Francis does too.

Curse of knowledge

Once we know something, we assume everyone else knows it too.
Example: Alice is a teacher and struggles to understand the perspective of her new students.

Spotlight effect

We overestimate how much attention people are paying to our behavior and appearance.
Example: Sally is worried about whether everyone's going to notice her new ice cream T-shirt.

Tool | Mind Maps and Social-Identity Maps

What's a good way to organize your ideas? At first blush, you might consider just listing your ideas:

1. Idea
2. Idea
3. Idea
4. Idea

… and so on. Or, maybe a grid to hold all your ideas (like a spreadsheet)?

	1	2	3
A	idea	idea	idea
B	idea	idea	idea
C	idea	idea	idea
D	idea	idea	idea
E	idea	idea	idea
F	idea	idea	idea
G	idea	idea	idea
H	idea	idea	idea
I	idea	idea	idea
J	idea	idea	idea
K	idea	idea	idea
L	idea	idea	idea
M	idea	idea	idea

This seems logical right? But, let's take a look at our simplifying assumptions. What deductive assumptions are we making? That all ideas are equal in importance, and that they all stand alone and share no connections.

What if we used our visual-hierarchy rules to distinguish some of these ideas? What if we were to arrange them using visual and spatial weight, which would then distinguish not only the significance of each idea but also the relationships between them? The result would be a mind map like the one opposite.

By taking advantage of visual hierarchy, we can use the size of an idea to group various different ideas into themes. We start out with a central topic, then the first main themes that connect to that topic spider out from the central hub. Radiating out from those points follow further ideas, and then the smaller details connected to those. We can also use all manner of art elements to enhance this visual hierarchy—for example, using solid and dashed lines to differentiate between direct and tangential relationships. Mapping out ideas in this way allows you to carry out an internal dialogue. As you generate fresh ideas, you'll find yourself drawing out connections and grouping ideas together as relationships surface.

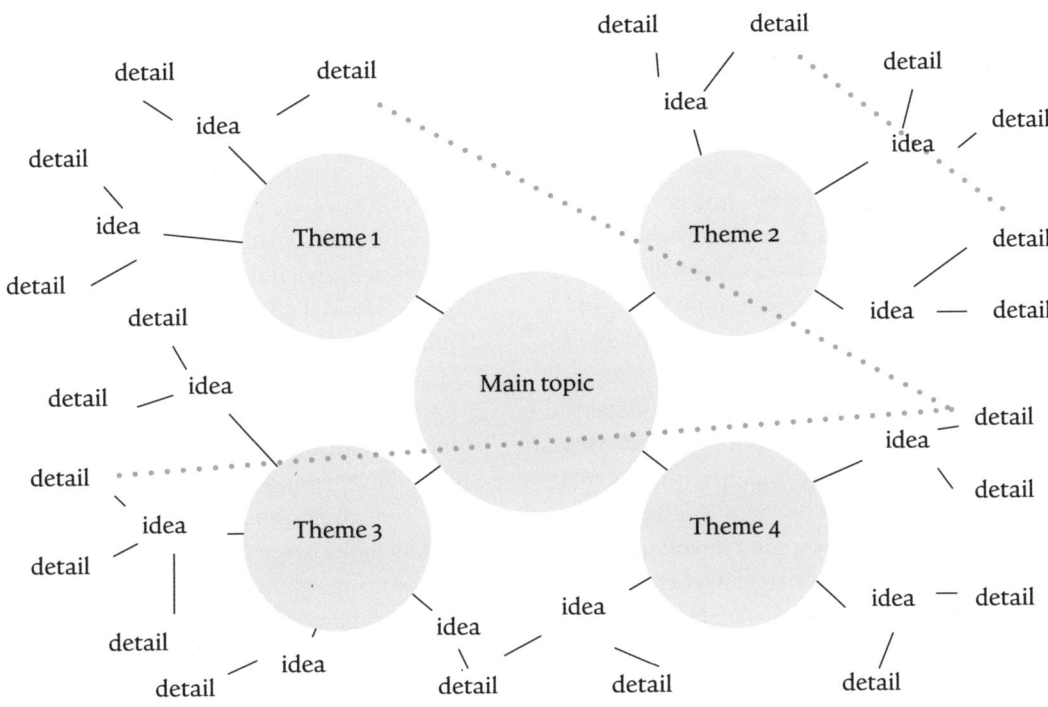

Quick tips for mind mapping

When preparing a mind map it's useful to be aware of a few things. First of all, keep your intention in mind. What are you trying to get out of the map? Organizing your themes? Exploring an idea or just recording a thought diary? Think too about expanding your view—is the thing you're focusing on a bug or is it a symptom of a larger issue? Relate your ideas to personal experiences to tap into your episodic memory, and describe your emotions to tap into subjective or subconscious thoughts. Think about the context of the idea.

Ask yourself questions as you work. A mind map is an individual narrative technique; you're having a dialogue with yourself, so tell stories, invoke metaphors, and make cultural references. Recall applicable childhood memories and other personal experiences, or refer to other media that relate to the topic, such as movies, books, or songs. Pay attention to your dreams and daydreams. Value intuitive leaps. Mind maps also benefit from quantity (greater size and increased density of ideas). Creativity strives for fluency (depth of ideas) and flexibility (breadth of ideas). Work on a bigger canvas if needed. You'll have more fun if you play with the form and presentation, adding color, drawing diagrams, attaching photos and media clippings, and so on. And finally, develop your personal mapping style—use humor where needed, and let your personality show.

Social-identity maps

We've already talked about positionality as an aspect of cognitive empathy. A social-identity map is a specialized form of mind map, designed to illustrate positionality. It was devised by Danielle Jacobson at the University of Toronto's School of Public Health.

The central concept is you, and the first themes are facets of your social identity. Note that the facets shown here are examples from Danielle, so feel free to map those that are significant to you. This first tier can be difficult because what one person finds significant to their identity may not be the same for another. For the second tier, think about how each theme impacts your life. Then, for the outer boundary, think about how that impact contributes to your emotional state. How does it make you feel or view the world?

As an exercise in cognitive empathy, can you imagine filling this out for another person? What would they write down? If you encountered that other person, and had them complete the map, how close do you think it would be to your version—would their map validate your hypothesis or illuminate a difference in how you both see the world?

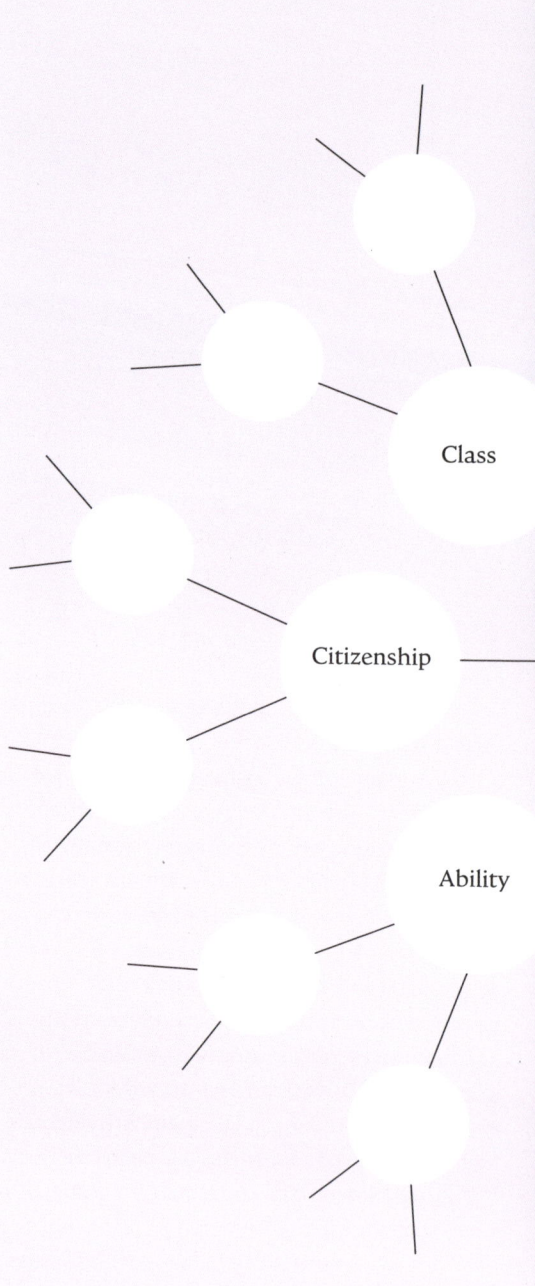

  Download the template and then post your social-identity map by using the QR codes.

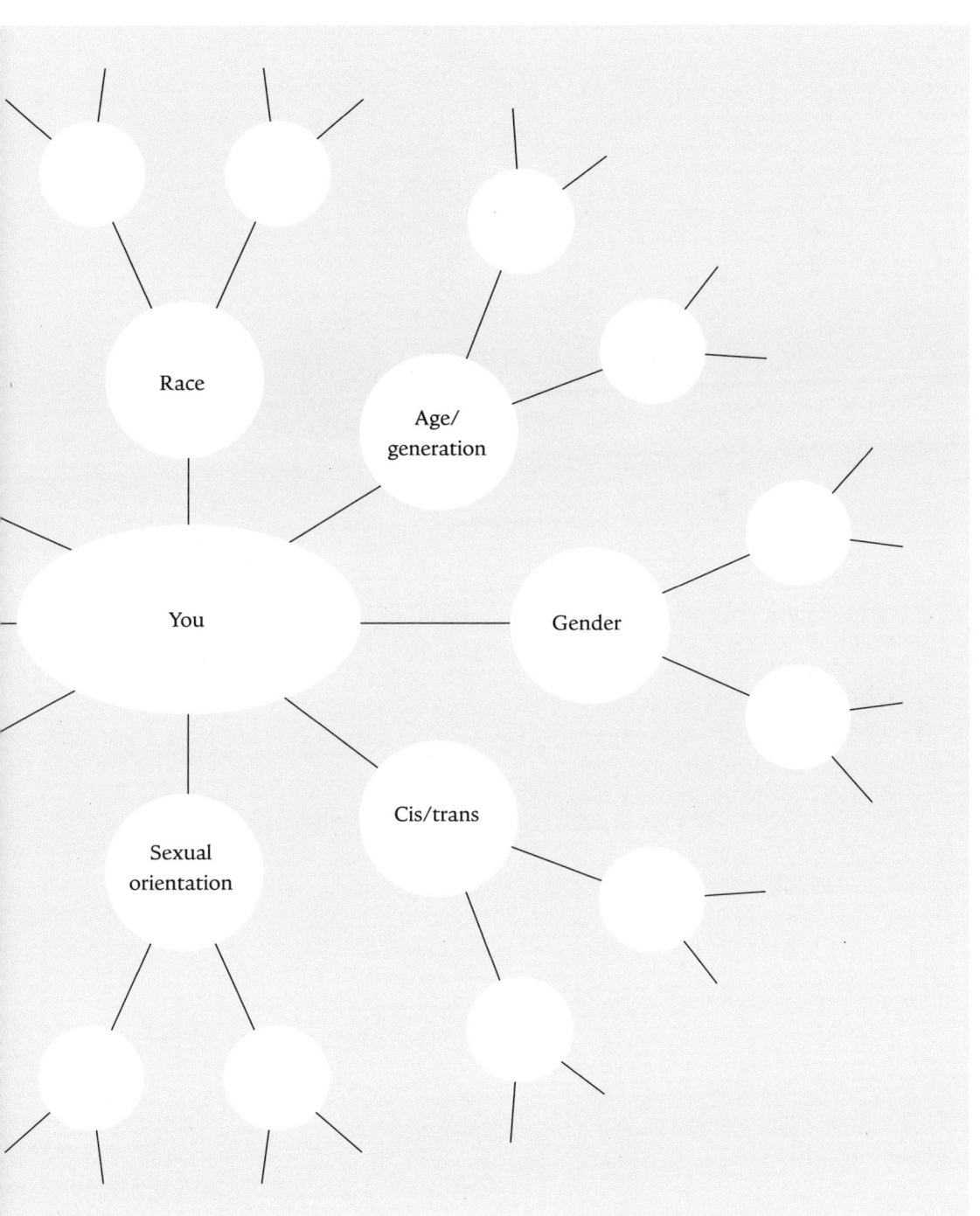

TOOL: MIND MAPS AND SOCIAL-IDENTITY MAPS | 137

3.4 | Emotional Empathy

While cognitive empathy is one of perspective-taking, emotional empathy— also known as "affective empathy"—is the ability to share and understand the emotions of another person. Within emotional empathy, you feel what other people feel, almost as if their emotions were contagious. This depends on your mirror neurons, which are cells in the brain that fire when you sense another's emotional state, "mirroring" or echoing a similar state in your own mind. Emotional empathy keys us in to the inner emotions of others and it's a fundamental component for many professions, from designers to salespeople, musicians to dramatists, and healthcare workers to first responders.

Emotional empathy consists of three components. The first is feeling the same emotion as another. The second is personal distress, where one has their own feelings of stress in light of another's situation. The third, much-studied, component is compassion, where you feel concern for another's plight. All three components help contribute to your ability to mirror the emotional state of someone else and experience something of their life.

While having an attuned ability to mirror emotions demonstrated by another can be an advantage in many professions, we also have to be cautious not to overstate this ability. Remember, due to positionality, these emotions are not exact mirrors. Emotional empathy is gained through remembering a similar experience in one's own life for reference.

For example, if you encounter someone who's just lost their job, you might perhaps be able to draw on your own experience of having been demoted. You can use that lived experience to access your disappointment at losing work and the anxiety you felt about your financial future. However, the two events are not exactly the same. If you were demoted because of a mistake you made, and the other person was fired due to their ethnicity, the context doesn't frame the emotions in the same way, nor how they came to be. So by saying you empathize with their situation, you may not be entirely correct, and doing so might even belittle their experiences. Recognize the limits of emotional empathy. It can help you understand people, but it doesn't make your outlook on their situation identical to theirs. The closer your experience matches another's, the more accurate your emotional context will be, but when the lived experience or positionality of the customer is far from yours, it can be better to involve them as a stakeholder within your design process rather than simply trying to replicate their emotional state. As a collaborator, they'll contribute their unique outlook and experiences to the product. Your role then is to use your empathy and design skills to help translate their needs into an actionable plan and viable product.

Emotional empathy can have other downsides as well, especially for those in healthcare and first-responder situations. If you experience overwhelming distress when engaging with a patient, that can leave you incapable of doing your job. Remaining rational enough to assess a patient's situation and apply your expertise is key. This is why doctors sometimes appear to have developed a sense of detachment when working in highly triaged situations. A patient may appreciate that her doctor is concerned, but she still expects professional and competent treatment, not someone who cries alongside her.

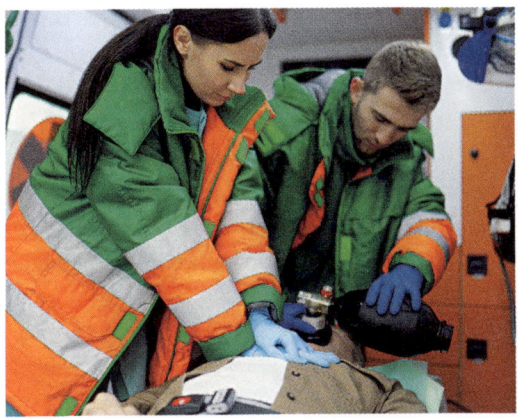

ABOVE: Emotional empathy can be a great asset, but for designers as well as first responders, it can be problematic if we feel so deeply that we're unable to respond to a situation or perform our duties.

Emotional empathy is also, understandably, a fundamental skill for people working in the performing arts. Whether they're opera singers, musicians, actresses, or dancers, all performers are trained to tap into the motivations and emotional states of a character and then channel that into their work. In design practice, emotional empathy can help break down the barriers put up by a wary audience, and draw you into a situation in which you are moved to help. Luckily, there are method-acting techniques that we can borrow from, designed specifically to develop emotional-empathy skills, so we'll look at a few of those next.

"Our human compassion binds us the one to the other—not in pity or patronisingly, but as human beings who have learnt how to turn our common suffering into hope for the future."
Nelson Mandela

3.4 EMOTIONAL EMPATHY | 139

Tool | Method Acting for Design

Method acting is a school of acting popularized by Lee Strasberg, a Polish-American actor, director, and teacher, in the mid-20th century.

Inspired by the "System" devised by Moscow Art Theater director Konstantin Stanislavski in Russia, Strasberg, along with Harold Clurman and Cheryl Crawford, formed the Group Theatre in New York in 1931 and began adapting Stanislavski's techniques. Strasberg eventually trademarked his own interpretation of these techniques as "The Method," which would go on to be employed by many famous actors and coaches, from Elia Kazan, Sanford Meisner, and Stella Adler, through to Robert De Niro, Daniel Day-Lewis, and Forest Whitaker.

Method acting trains actors to use their own emotional, physical, and mental experiences in the creation of the character they play. The techniques require them to tap into their personal experiences, reliving them, as a way to invest their character with emotional authenticity. As a style of performance, method acting encourages performers to embody their characters not just onstage but also in everyday life. Embodying the character involves deep research into their psychology and emotional state, and often living life "as the character."

BELOW: Judi Dench and Daniel Day-Lewis performing in a 1989 production of *Hamlet,* on the Olivier stage at London's National Theatre.

ACTIVE MEMORY EVENT

Affective memory

Write down a personal experience you remember vividly:

_____

Write down the emotions felt. Be specific:

Sense memories

Sights:

Sounds:

Smells:

Tastes:

Textures:

Let's take a look at a few exercises that will help build your emotional empathy.

EXERCISE 1: AFFECTIVE MEMORY

Method acting is known for its application of "affective memory," which requires an actor to call upon an experience from their own life—to recapture and relive it, then deploy it in the service of portraying their character. For example, if an actor is playing a character who has just accomplished something important within the story, the actor will try to remember a moment in their own life when they felt proud of a significant achievement.

Think back to a time in your life when you might have felt great pride—perhaps attending your graduation ceremony. Can you remember how you felt, the emotions of having just accomplished something that felt so hard at the beginning? Can you recall the struggles you might have overcome? Perhaps your loved ones were in the audience, there to cheer you on and celebrate with you. Your emotions may have been mixed—proud of the work you'd completed and thankful for the results, but at the same time, humbled by the thought of the sacrifices your friends and family had had to make to put you into that position. What sights and sounds do you recall? The perfumes worn by people sitting close to you? What did your graduation robe and satin stole feel like—smooth between your fingers, cool to the touch, contrasting perhaps with a slightly sweaty palm?

How about a time when you felt truly connected to someone? Maybe it was an exhilarating win for a sports team, or your wedding. Or what about a sense of loss: a friend who may have moved to another town, or the death of a family member? By remembering emotional moments in your life, you reconnect to your humanity and the sorts of moments many of us share.

EXERCISE 2: SENSE MEMORY
A sense-memory exercise asks you to remember and feel all five senses of an experience in preparation for duplicating an emotion for a scene. For example, think back to a cherished childhood memory, perhaps receiving a favorite toy as a birthday gift. Can you remember the smell of the packaging as you ripped the paper off? Hear the crinkling rustle of cellophane as you revealed the toy? What about the toy itself—how did it feel or smell? Hard plastic with a touch of machine oil, or a fabric-softener scent of furry fluff on a stuffed animal? Was there a taste from earlier in the day, such as a celebratory piece of cake at lunch? This experience probably involved a mixture of emotions: giddy anticipation, joy, and surprise all mixed into one.

For an actor, a sense memory like this can be used as a tool for recalling the emotions needed for their character. For any creative practitioner, it is a way to connect with an audience. Try to remember a few key experiences so that you then have them on recall. This will heighten your ability to identify the same emotions in others.

EXERCISE 3: LIVING AS THE CHARACTER
Coming from the idiom, "You don't really know a person until you've walked a mile in their shoes," I like to refer to this exercise as a "moccasin," designed to explore the life of another person. This is a method-acting technique in which you try to live the life of the character, understanding them through full immersion.

For example, if you wanted to get under the skin of a character who's bound to a wheelchair, you'd rent a wheelchair from a medical supply store and try to live with it for a week. You'd quickly discover that all the cabinets and drawers in your home were no longer appropriate for you, and that the shelves in the grocery store were no longer within your reach. You'd probably also experience frustration at all the detours you'd have to make just to get where you'd like to go, whether that meant taking a different form of transportation or just locating a wider entrance into your school or work building. Experiencing someone else's physical and emotional journey instantly builds empathy and understanding, which is why so many method actors have carried out similar preparation for their roles. Robert De Niro, for example, worked as a cabbie in New York before *Taxi Driver*, and Hilary Swank lived and trained as a boxer to play one in *Million Dollar Baby*.

LIMITATIONS OF THESE EXERCISES
While practicing and utilizing these exercises is a great way to help you attune your emotions to others, I also want to stress that they are not a duplicate of other people's experiences. For example, in the wheelchair example, understanding the emotions of being confined to a wheelchair will help you identify with a wheelchair user, but that person may have spent their lifetime in a wheelchair, whereas you've

only used one for a week. Your initial frustration at your drawers being too deep or your cabinets too high might have been a major inconvenience for them 15 years ago. They may have already resolved that issue by lowering their drawers and shelves—or they may have never even experienced it in the first place because their family remodeled the kitchen before their arrival home from the hospital. On the other hand, perhaps something that you found slightly inconvenient, such as transportation, continues to be a major headache for them due to differences in circumstances.

Again, positionality differences can limit the accuracy of your emotional empathy. You cannot put complete faith in a specific method-acting exercise then generalize about an audience or group of people. It would be presumptuous to assume you knew what all wheelchair users go through because you spent a week in one. This group is not monolithic—different users come by this piece of equipment through different paths (injury versus lifelong disability). Nor are their needs universal. A person with cerebral palsy may need a powered wheelchair, whereas an athlete with a leg cast can get by with a mechanical, non-powered chair. Like all tools, this one has its benefits and disadvantages. Just learn to use method-acting exercises for what they bring to your process, but remain mindful of their limits.

ABOVE: Daniel Day-Lewis employs method-acting techniques to tranform his appearance, mannerisms, and speech in (top to bottom) *Gangs of New York*, *Lincoln*, and *Phantom Thread*.

  Download the template and then post the results of this and other method acting exercises by using the QR codes.

TOOL: METHOD ACTING FOR DESIGN | 143

3.5 | Compassionate Concern

The combination of cognitive and emotional empathy leads to the third type of empathy, empathetic or compassionate concern, where the combination of both understanding someone's plight and emotionally mirroring them, spontaneously moves you to help that individual. This last type can be tricky, which is why compassion is often a topic of investigation for social scientists.

In today's media-saturated environments, where global suffering and injustices are constantly on the news and social media is ever-present, it can be easy to become inured to the plight of others. People tend to put up mental or emotional blocks to cut off the suffering of others because overwhelming media coverage leaves them feeling unable to respond in a meaningful way. It would be hard, for example, to drop everything and travel overseas to help with hurricane relief work. How do you maintain a balance of the two empathy types in order to transition into compassionate concern? It can be tricky.

Holding both the cognitive empathy mindset (understanding another's situation with respect to their positionality) in the same hand as emotional empathy (where you inherit their emotions as your own) is not an easy task. The former is analytical in nature and can cause us to detach from the situation; the latter is emotive and draws us into the fray. However, instead of moving us to action, an excess of emotional empathy can also paralyze us into inaction.

The goal then is to try and balance the first two empathy types with remembering your purpose: to create a product that will improve another person's life and enhance their well-being. This is our contribution as designers and creative individuals—to offer a creative leap or output that you hope improves upon the situation, and to bring that to the community as a discussion point for engagement.

If you feel overwhelmed, remember to set boundaries for your activities (especially when practicing method-acting techniques), and learn to rely on a strong support network of friends, family, and other associates to keep you grounded. If you feel yourself overanalyzing positionality, or becoming stifled by "what if's," spend some time employing "moccasin" work (see p. 142) or conversing with your audience, to reconnect emotionally. Don't forget the objective—you're trying to understand the plight of another, cognitively and emotionally, in order to draw inspiration from it and be moved to create for your audience.

Actively listening to your subject and finding a connection face to face, in their environment, helps build your emotional empathy. Being genuinely interested in another person's life, asking questions that are pertinent to what they've shown or revealed to you will foster stronger rapport and emotional connection. Then, after a face-to-face, take the time to reflect on what you've seen and heard. Make notes, then move into your cognitive empathy state. Why do you think they spoke or reacted in certain ways? What in their life's history do you know of that might have influenced their behavior, and how does that compare to events in your own life? Try creating a social-identity map (see p. 136), inserting the factors and needs that you've noticed. By balancing your frame of reference, first as a participant living in your audience's life, and then as a researcher, analyze how points of view were formed and why certain

ABOVE: Industrial design, computer science, and engineering students on a community walk along with the neighborhood's civic association president.

emotions may have developed. Now comes the creative leap: compassionate concern. Can you imagine a better future for their community? What can you create to improve the situation?

I've occasionally encountered researchers and designers who've taken a counterpoint view on empathy, insisting that it has no role in design. They will point to a specific case study, or the old adage that "Customers don't know what they want," or the Henry Ford quote, "If I asked a customer what they wanted, they would say a 'faster horse.'" But for every bad outcome I've seen, it was never an empathetic mindset that was to blame, but a faulty design process. For example, a designer being so moved to action that they placed blind faith in their own method acting without analysis or an understanding of positionality. Or a designer being so inspired by a situation that they fixated on the very first idea they conceived to solve it, mistaking it for some kind of divine intervention. This speaks to inexperience in the design process and naïvety with other design phases, such as testing. In a rush to help, that designer never stopped to circle back more analytically with their stakeholders, or to evaluate their design to ascertain whether, in addition to it being good for their customer, it was appropriate for the people involved in bringing it to fruition. An empathetic mindset is critical to your design process—as important as creative and critical thinking. Failing to recognize its role alongside other design phases is a disservice to the practice of product design.

Tool | Analogous Research

To help develop compassionate concern, you need to combine your cognitive understanding of what you want to design and for whom, together with your emotional understanding of the context at hand.

One tool that helps cultivate both cognitive and emotional empathy in order to trigger compassionate concern is an analogous research study. To conduct such a study, you look for a relevant example from a different world than the one that you're designing for, then observe or participate in that activity so that you have a proxy for the real context. While an immersive, "living as the character" method-acting technique will help develop your emotional empathy, an analogous study helps give you a little distance from your subject and triggers your analytical mind to compare the differences between the study and the true context of the design. It can also be very handy when the original environment is challenging to design for, perhaps due to logistical difficulties arising from regulations or availability of location.

For example, if you were tasked with designing a product for use in hospital operating rooms, observing a pit crew at a NASCAR race could offer useful insights. Both scenarios have an individual at the heart of the procedure, and in both cases, time is critical to success. Each scenario is also characterized by intense teamwork carried out by people with different specialties, so communication between team members is crucial. How might you use the NASCAR analogous study to learn about the hospital context? Are there any differences in perspective (cognitive empathy) or emotional states (emotional empathy) among the participants? What fresh inspiration can be gleaned?

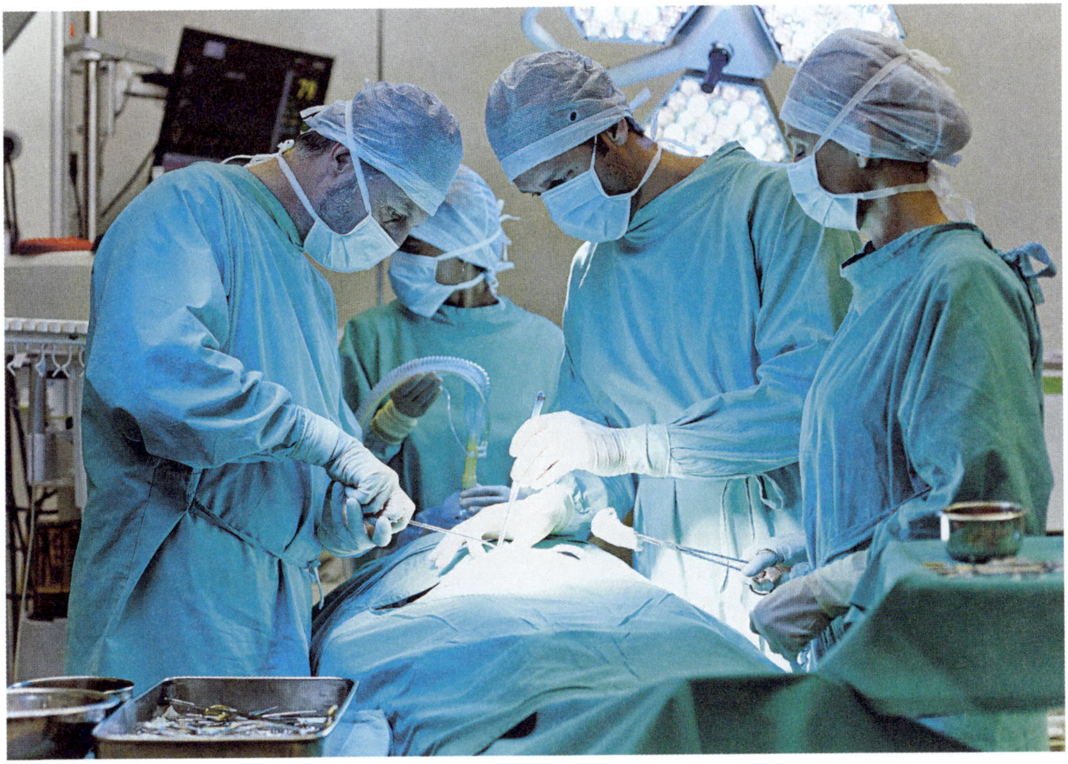

When conducting analogous research it's best to find examples from worlds that share similar guiding principles. These are not simulations of the original context but they are authentic—they make connections that bring familiar experiences into unfamiliar territory. It's also usually easier to perform analogous research using several smaller examples, sharing the range of different qualities between them as a combined proxy, rather than searching for one big, all-encompassing example.

OPPOSITE: NASCAR pit crews display teamwork and distinct roles in order to quickly and accurately perform tasks on a car mid-race.

ABOVE: An operating room, where a similar display of teamwork employing coordinated individual duties occurs. What might we learn by comparing and contrasting the two?

Download the template and then post your results by using the QR codes.

TOOL: ANALOGOUS RESEARCH | 147

3.6 | Narratives and Interviewing

Why are narratives so important when it comes to understanding people? Narratives, or stories, are highly memorable and they are the way humans naturally tend to describe their lives in order to relate to others. Stories can be informative, such as a news story, spread through both visual and verbal channels, and they can travel quickly (through media). They can be used for positive ends—to inform or to highlight social tension or injustices, for example—but at the oppposite end of the spectrum, they can misinform and sow division. The very potency of storytelling is why those in the media bear such a great responsibility to avoid distrust in the medium and to represent the stories they tell faithfully and factually.

There are many examples of effective and memorable narratives. One known by many is the story of Noah's Ark and the Great Flood, which features in varying forms in the texts of Christianity, Judaism, and Islam. Some stories are passed down through oral tradition, like Homer's *Odyssey*. Others are brought to life through novels like Charles Dickens's *A Tale of Two Cities*, plays like Arthur Miller's *The Crucible*, or movies like those produced by the Marvel Cinematic Universe.

In the "Frames of Meaning" section in Chapter 1, we looked at Jerome Bruner's intentional states (see p. 44), then used empathy maps to more accurately describe them as say-do-think-feel intentional states, which I will shorten here to "intentional states." As collections of these intentional states, narratives provide a means for us to share our values and beliefs with others. Narratives sustain and transmit culture, which can be defined as the way a group of people gives meaning to the world around them.

These cultural vessels are sequential (they have a beginning, middle, and end), define what is normal and acceptable within the culture, and they often showcase dramatic elements that point to tension or unresolved human needs within a society.

When you converse with people, they reveal their values, attitudes, and behaviors through the stories they tell you. Not only do they inform you of how they'd act in a given situation, they also reveal how others might support or oppose their actions. They help describe the unspoken rules of their culture. This can be seen in the following few sentences:

> So then I left for college. My family, immigrants from Thailand, had never gone to college and I was the first to go. They were so worried about me—that I'd lose my roots and lose my way in the big city—but here I am. I go back home at the end of every term.

Something that short and yet that descriptive communicates so much about a culture's needs and value system. These two sentences form a mini-story, which already has a story structure: an exposition (beginning), a dramatic point (rising action leading to climax), and a denouement (ending). This short text reveals a great deal of information: established tradition versus new behavior; a clash of cultures, between an immigrant family (Thai) and their adopted country; a simple, rural lifestyle contrasted with metropolitan, progressive city life. A journey is implied too: The student leaves to go to a new place, with a resolution—the fact that he always returns home.

Not everything is resolved, though, merely implied. The student states that he's not forgotten his family or his culture, and that college hasn't changed him. But is that true? How would you be able to verify this assertion of values? You'd have to assess his intentional states and see if they contradicted his story. If he were to bring home a non-Thai partner, for example, or miss a holiday break and remain in the city, then you'd have evidence of a conflict in an intentional state and you could claim that the student had indeed been changed by his college experience. You could also interview his family members back home to see if their views of the student aligned with or contradicted his story.

Remember, it's the contradictions in intentional states that point to unresolved needs and tension within a culture or community. In this hypothetical case, it's the need for an immigrant population to acquire an education that allows them to join a modern workforce, while also maintaining their unique culture and celebrating it within the larger society. Contradictions like these are what present opportunities for designers to have an impact.

Stories like the one here are gold mines of information, both explicit and implicit, so they are what designers use to identify and organize human needs into actionable designs. When engaging with potential customers and stakeholders, you can unlock information about your audience and the context by listening for three different types of stories: contradiction, normative, and success and failure. We'll look at these next.

BELOW: An ethnographer interviews a local woman while on a research study on the Salween River in Thailand.

1. Contradiction: The first story type is self-explanatory. Watch what people do, and listen to what they say about what they do. Take note of their stated actions and see if they match up with their behavior. Inconsistencies within behavior, context, or stories are a sure sign that needs are not being met. The speaker's need to resolve these contradictions provides an opportunity for new products and services. Often, you'll be able to detect rationalization of the contradiction—cognitive dissonance, which is characterized by holding two competing values at the same time (see pp. 44–45).

2. Normative: These are the stories in which you often hear references to identification within a group. "This is the way we've always done it," or "This is how the union operates, it's business 101." Normative stories define boundaries of behavior and are revealed in standards, morals, cover-ups, shame felt when going against the grain, "shoulds" and "should nots." Norms point to "other" optional behaviors, mediating strategies, or people's desire to conform to the norm. People feel the need to be perceived as normal by the rest of the group, so if their behavior contradicts the norm, they'll feel compelled to find a story that explains or mitigates it.

3. Success and failure: We've probably all heard one or two of these stories before. These are the tales of the time when your team won the state championship, or didn't make the playoffs. The trials and tribulations of raising kids, or the triumphs of effective parenting. A story of this kind will often start with, "Let me tell you about this one time when…" and go from there. Success and failure stories describe events and contain the implicit rules that govern people's lives—what values they promote in order to succeed (or ignore and therefore fail). When you hear these, try to understand the "why" of the story's situation, and the context and factors that led to the outcome.

So how do you approach someone and engage them in a conversation that will result in a useful story? In the social sciences and communications fields, specific techniques are used to draw these types of stories out of people, in a neutral manner, without the interviewer's own bias muddying the data. They work quickly too, with interviewers sometimes able to get the stories they're after within an hour or so. These techniques, often referred to as "open-ended, semi-structured" interviews, or ethnographic interviews, are used by social researchers, designers, and law enforcement alike to quickly capture the "raw data" of a situation, which is then analyzed after the session. The section that follows describe how to conduct interviews like this for your own work. The techniques do take practice, but the more interviews you conduct, the more competently you'll be able to facilitate them in future.

OPPOSITE: American journalist Terry Gross hosting *Fresh Air*, an interview-based radio show produced in Philadelphia and distributed in the US by NPR.

> "Don't ever diminish the power of words. Words move hearts and hearts move limbs."
> Islamic scholar
> Hamza Yusuf

Tool | The Open-Ended, Semi-Structured Interview

Interviewing techniques are commonly practiced in communication (journalism), social sciences (cultural anthropology, sociology), and law enforcement. The "open-ended, semi-structured interview" is an ethnographic interviewing technique taught by Dev Patnaik and Michael Barry as part of the Stanford design program.

This a particularly good cornerstone for learning about the everyday lives of people. It's a strong formative research technique, used to gain insight into an area and clarify an issue you're trying to address. (When you already have a prototype or product built and are trying to evaluate its effectiveness, there are other techniques for engaging with people. Known as "summative research" methods, these might include web analytics [online reviews], evaluation matrices, and customer satisfaction questionnaires.) Because of the nature of the open-ended, semi-structured interview—used to explore people's lives and value systems, so that you can understand their needs—an empathetic, genuinely curious mindset is best. You'll later reflect upon and organize those needs in order to identify an opportunity for design. Unlike a satisfaction survey, where you're trying to rank quantitatively someone's opinion on an existing product, you're using this technique to capture their intentional states (pp. 44–45) and explore their lives, which is more qualitative in nature.

This technique is very conversational and, though it has some structure, there's a little bit of improvising and being in the moment. You'll still prepare an interview guide (a list of questions you'd like to ask), but instead of working by rote through a series of multiple-

The interview structure

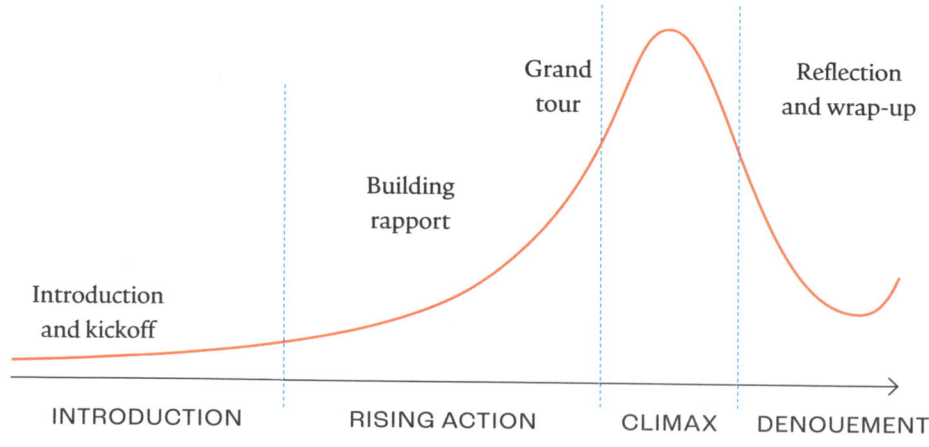

152 | CHAPTER 3: EMPATHY

choice questions (quantitative survey), think of yourself more like a DJ at a party, reading and reacting to the person in front of you, selecting questions (then asking follow-up questions) on the basis of the answers they provide. There will be opportune times to ask specific types of questions, based on their responses and the conversational flow. When you first meet a stranger, you're not likely to trust them with personal information or reveal your inner values or personal beliefs. In order to learn more about people, you'll have to conduct this interview personably, and in a way that builds trust.

The interview structure

Much like the stories and intentional states we're wanting to record, the interview itself is built around a story structure. The conversation will run through four main phases: an introduction, rising action (building rapport, followed by a "grand tour"), a climax, then a conclusion.

INTRODUCTION AND KICKOFF

These interviews are best conducted in the field, where the participant most feels at ease, and where they conduct the activities that you're interested in. The goal is to gain insights into their regular behaviors; if you invited them into a lab with mirrored walls, they'd know they were being studied and would find it hard to act naturally. Set up a comfortable place for the interview and begin by explaining your purpose, telling them what you'll be doing with the information they provide. This is the time to review a consent form, which is a document signed by the participant giving permission to be recorded, and establishing the degree of privacy they're comfortable with, such as whether they're willing to be recorded, and how much of those recordings they'd accept being made public. Let them know that their input is important and, as an ethical designer, reassure them that the consent form is an agreement on how you will treat their information.

Question types: In this early phase, there is the natural defensiveness that someone being interviewed feels toward an interviewer. They're trying to understand the purpose—what their words or the recording will be used for. To help ease this tension, questions that you can ask would be:

1. Introduction: "Why don't you start off by telling me a little about yourself?" Or, "What do you do here day to day?"
2. Sequence: "Can you walk me through a typical shift when you get in, in the morning?" "What do you do first? ... and then next?"
3. Specific example: "Let's start with yesterday. What do you usually do every day when you wake up? Can you think of anything non-routine?"

RISING ACTION: BUILDING RAPPORT

Once the introduction has been completed, you'll again want to ease defensiveness with some reassurance about your responsibility as an ethical researcher as you embark on the next phase of the interview: rising action. You can start to build on your initial questions with some general, concrete inquiries. People like answering questions they know the answer to—it helps them feel more confident—so asking detail-oriented or overview questions are easy ones to start with. Usually these questions will lead the participant to talk about their experiences, which can lead into a story about their work or life. It's important to just be attentive, and listen. Fight any urge to interject.

Remember, this technique is focused on learning about their lives, not about fixing your problems. You'll know you're doing this right when your cognitive-empathy mindset is firing and you're talking far less than your subject—perhaps only to ask the next question, or to clarify a detail. You'll also be able to gauge whether you're building rapport from the level of interest they show in the conversation, and from what they choose to share with you. If they lead into a personal story, give them the space to tell it, remain curious, and let them finish.

Question types: Building rapport means asking general, concrete questions, which establish both detail and overview:
1. Task-based or organizational: "How many people work here?" Or, "What are the tools you use most often to do your job/activity?"
2. Exhaustive list: "What are all the drinks that you offer here?" Or, "Can you show me all the ingredients that make up an Old-Fashioned cocktail?"
3. Quantity: "How many people would fit on your team?"

RISING ACTION: GRAND TOUR
This is the second part of the rising-action phase. As your participant acclimates to you, you can begin to explore their environment. Ask them for a guided tour of the setting. Ask questions, and act out scenarios (as if method acting their activities). Rapport and trust should be building, and at this point you can begin to ask questions about scenarios the participant has been in.

Oftentimes, while acting out an activity or asking a question about a task, an interviewee will open up with a story that may surprise you. Or, they may say something that just "hits different." If that happens, be sure to note down the exact words because these "Aha!" moments are important points of connection. Often, when you later recall the words, it will reveal an insight about that person, an interpretative statement describing an implicit belief or abstract need. Questions at this point can get personal, but that will be dependent on the conversation itself. Either way, if you experience a revealing moment like this, that's a sign that the interview has reached its climax, and an indication to begin wrapping up.

Question types: Grand tour questions can investigate the environment and the scenario you're acting out. While in this phase, if the opportunity presents itself, you can dive into questions that are more personal:
1. Suggestive opinion: "Some people say social media connects us, while others say it divides us. What are your thoughts on the subject?"
2. Native language (when you hear terms that sound foreign to you in context): "Why do you call this 'Agent Q's workshop?'"
3. Clarification (often when you hear a colloquial phrase): "When you say he was 'well put together', what does that mean?"
4. Empathetic: "Can you show me how I should do xyz activity?"
5. Naïve outsider perspective: "Let's say that I wanted to start playing pickleball. How would I go about doing that?"

REFLECTION AND WRAP-UP
After the climax of the storytelling, the interview enters a reflective stage. You can inquire about thoughts and feelings, as well as comparisons with others. At this point, you should also have engendered enough trust between you to be able to ask more personal questions. At the end of the

interview, you can explore more abstract thoughts and feelings. A wrap-up involves letting the participant know that the interview is coming to a close, usually by telling them how much time is left: "We've got about ten more minutes," or "Thanks again for the opportunity to talk to you today." An interviewee will often come forth with additional information once you've signalled that the end is in sight. They may want to clarify something they said earlier, or if sharing a connected moment, they may want to elaborate on a particular topic.

Question types: Reflection and wrap-up questions can be about thoughts and feelings, both concrete and more abstract. Questions in this phase are intended to spur reflection:
1. Changes over time: "How are things different from how they were a few years ago, before the storms?"
2. Projection: "Do others feel the same way as you, or do they feel differently?"
3. Why (use with great care): "Why did (or didn't) you do xyz?"

Advice for conducting interviews

Though I mentioned that conducting an interview is like being a DJ, where you're reading the room and selecting the right song from the right genre at the right time, this doesn't mean that the interview is complete improvisation. The improvisational aspects come with being in the moment, present, attentive, and reactive, able to pose the right type of question at the right time because information you hadn't anticipated has arisen. Doing some research on your participant beforehand—their role, the activity they perform, and the environment in which they work—is encouraged. Creating an interview guide, consisting of five or six questions of each type, in anticipation of what you think may happen, is also a great way to prepare. It will make you feel more confident when in the field. Don't cling blindly to the guide, just have it in your back pocket for reference—it's more important to read and react to your participant. You will sense the flow of the conversation, and if someone is more apprehensive, you may have to move slower through the phases, or spend longer in an early phase to ease their defensiveness. It's critical to remember that you're using empathy to learn about others.

After the interview, be sure to reflect on what you heard. Investigate any particularly insightful quotes that you might have jotted down; describe the needs and intentional states that you observed, perhaps analyzing them using an empathy map, and take the time to review any stories they may have shared with you.

Download the template and then post your results by using the QR codes.

Tool | Journey Mapping

This tool, known as a journey map, comes from Jakob Nielsen and Don Norman, who formed NN/g, a user-experience and computer-user interface consulting firm. The journey map provides an excellent template for charting your AEIOU observations (see pp. 88–89) and making sense of human activity.

While the tool was designed to study how computer users accomplish tasks, it can be applied to any activity where interaction occurs (between objects, people, and environments). Activities like ordering at a restaurant or bringing in a vehicle for service are a good fit for journey mapping; it helps to identify opportunities to improve upon a customer's experience as well as the services designed for them.

Zones

As seen in the template opposite, a journey map is constructed in three different zones, addressing the subject of the map, their experiences, and your insights.

ZONE A: THE LENS
This identifies the user or customer and what they are attempting to do.

Components: Upon carrying out enough observations, you will most likely have an aggregate understanding of the people trying to accomplish a task. For privacy reasons, you won't use an actual participant's identity; instead you use a representative "persona":

1. Persona photo: fictionalized visage, but based on your observations.

2. Scenario/goals and expectations: What's the activity that you're mapping? Describe the task at hand. What does the customer expect to accomplish and how? What determines a successful activity in this case?

ZONE B: THE EXPERIENCE
The middle of the template is the bulk of the recorded activity and this is where your AEIOU notes come in handy. What is the activity in step-by-step detail?

Components: Go through your notes and assemble any interactions, interesting insights, quotes, and pain or gain points. Look at the activity and go through it step by step:

3. Activity headings: Group your activity into phases that make sense. For example, in ordering coffee, you might have a section on pre-ordering, ordering, and receiving the order.

4. Step by step: Note down all the steps and interactions that you observed.

5. Photos and quotes: You can include any quotes, photographs, and video files that you recorded, especially those that you feel were either beneficial or caused problems.

6. Emotional arc: This is a line that charts the customer's emotional state (happy or sad, good fortune, ill fortune) on a vertical axis.

ZONE C: THE INSIGHTS
The bottom of the map charts the "Aha!" moments you observed. When interviewing and running through scenarios with your participant, surprising moments will correspond to a point in the activity and are often areas for improvement.

Components: What interesting observations might lead to a process or service improvement:

7. Opportunities for improvement: How might this part of the activity be redesigned, made more efficient, or obsoleted altogether? Would a reordering of steps create a different heading or a new journey entirely?

8. Internal ownership: Who is responsible for making sure that insights identified are addressed and that changes in the process are executed?

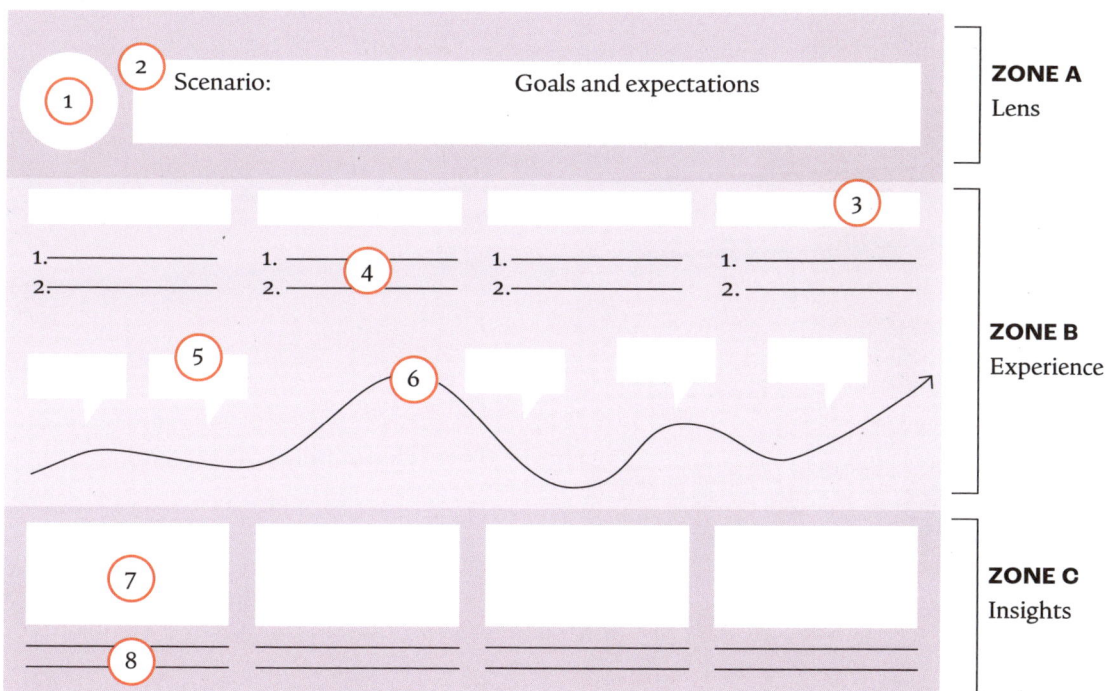

ABOVE: A journey map template from NN/g, which collects information about an activity or experience and presents it to an audience.

  Download the template and then post your journey map by using the QR codes.

TOOL: JOURNEY MAPPING | 157

4

CONTEXTUAL AWARENESS
Cultivating Your Designer's Sixth Sense

HOW DO YOU DETERMINE WHICH DESIGNS ARE GOOD OR BAD? UNDERSTANDING THE CUSTOMER, CONTEXT, AND MARKET IS YOUR SECRET WEAPON.

4.1 | An Interview with Christina Choi
4.2 | Expanding Your Frame of View
  Tool: Powers of 10
4.3 | Trends, Technology, and Culture
4.4 | Organizing Human Needs
  Tool: Common to Qualifier
  Tool: Affinity Mapping and 2x2 Matrices
4.5 | Timelines and Mapping
  Tool: 9 Windows
  Tool: Era Analysis—Material Culture
4.6 | Design and Business Strategy

4.1 | An Interview with Christina Choi

Design education across three contexts

Christina Choi is an educator and researcher, Conran Chair, and Head of Programme in Design Products at London's Royal College of Art (RCA). She's also a colleague from her days at Georgia Institute of Technology, and a good friend.

Christina completed her BFA and MA in industrial design in South Korea. Industry work led to a desire to explore technology and design. She completed a master's and doctorate at Georgia Tech in Atlanta. She's a much-published professor, contributing to journals and conferences, and a member of the editorial board of the *Journal of User Experience*. I was very happy to be able to sit down with her to discuss her design experiences.

Li: Christina, let's start off with your career. It's spanned design across three different countries: Korea, where you studied design; North America, where you did your graduate studies and were also a graduate coordinator and a professor; and now Europe as the head of a design products program. Tell me a bit more about your experiences across these three regions. What's the same? What's different? And do you think culture affects design and design education?

Choi: Absolutely. I'd say there are differences in focus related to making, practicality, and conceptual design. Korea is a good place to learn overall design skills and base design, to realize your idea, except for one thing—the physical making of things is a little bit less rigorous than in Europe or the US. For example, I recently finished a workshop with 16 students at the Korean Institute of Design Promotion. During the workshop I talked to them about Korean design education nowadays and asked them, "When you make your prototypes, do you send them out to the professional shop for graduate projects?" And they said, "Yes they have a shop, but they're not making things … still." There, making things is not their focus. Similarly, many of my RCA students from China strongly believe that some part of the production has to be done in China. It's cheap and fast, so they try not to make their own prototypes. I see their point because they'll say, "Well, when I get the job post-graduation, somebody there can make the perfect functional model. Why do I have to do it?" As an educator, what I'm asking of them is to deliver a certain level of functional model that they can test.

So anyway, back to my comparison. Korea is a good place to learn overall design skills and base design, but not necessarily for making things perfectly. For the US case? I can obviously say with my time at Georgia Tech that students work with engineering, make functional models, and spend time on research. When I was Graduate Program Director I was always telling prospective students, "When you graduate from Georgia Tech you're going to be really amazing at three things: you're going to be really good at research; you're going to feel really comfortable with working with multidisciplinary teams, especially engineers; and you'll understand many different disciplines, which is related to working with different people from different places."

I would also say that in the US, design is super practical. Think about the car, for example. I mean it's very comfy, still very practical, but might not necessarily … US car design doesn't make your heart pound. I'm sorry, that sounds too negative. It's maybe more generic. It can be very strong, very rigid, practical, but still, aesthetically sometimes, not very detailed. In the UK case, there's more emphasis on concept design. I've been here just under three years, so it may be too early to say that. But there's much more focus on concept design and then experimentation. I'm not talking about experiments in research but about exploring more radical concepts or approaches. For example, one student was designing a beautiful bench, and then he went to design a pattern for it. What he did was put oil in water, and then when the oil made a pattern on the surface, he took a picture of it and used it to cut into the metal to make a bench. It was beautiful.

For education, the UK is more pioneering in terms of new teaching and design methods. For design theory, a lot of influential designers are from Europe. For overall spirit, Korea—lots of incremental innovation. They always try to work better; incremental innovation, but higher quality and cheaper. So for the same money, if I paid it, Korean products would be high quality and reliable. I can really confidently say that. The US—practical again, but more attempts to add radical innovation. Steve Jobs, for example. There are a lot of controversial things about Jobs, but I still very much admire him trying to design something we'd never seen before. I love that. And then the UK—I've got to say, tradition. And they create social sustainability, they take action every day to make everything sustainably, and also try to think about new ways to teach. So all countries deal with students' mindsets.

Li: I was actually going to follow up with that, because I can see both the pros and cons of what you describe. On the one hand, if you say, "Well, Asia has the factories here, so why don't I just send it to the shop? I don't need to learn how to make it"—the negative, of course, is that you don't learn production. On the other hand, you do learn how to leverage other people. So I can see both good and bad. Same thing with the US. It is very practical, which means it's robust and very useful, but at the expense of either aesthetic or experimentation because you're leaning so much into the practicality. It's both plus and minus. In Europe, where you say it's more about methodology or experimentation, it's more art in that sense; the Royal College is more art-based than Georgia Tech. What do you think causes those differences?

Choi: So it's kind of—people, history, and resources. When I'm talking to you, I could talk all night because you give me so much food for thought! So for example, in the UK, the tradition I've been talking about—there is a leather makers' guild that wants to preserve the tradition and history of the craft, but it's much harder now to find this sort of work because fewer people are utilizing artisanal products like this nowadays. And when they do use traditional UK leather, the labor costs make it expensive, right? The product is beautiful, and there's still definitely work going on, but is this the way they will all survive? So they're dealing with history, resources, and also culture.

One other issue related to undergraduate students' mindsets: They are always "jumping to the solution" or implementation before understanding the problem. This is a design disease. It doesn't matter where it is—the UK, Korea, the US—they're jumping on the solution. It doesn't matter the diploma or major, either: design, mechanical engineering, ECE [electrical and computer engineering]; nobody does enough research for a solid base understanding to accurately define a design opportunity. They keep doing the "jump on the solution." I don't know why.

Li: Let's talk about that design methodology—the way that students approach design, with this natural tendency to just jump to a solution. What skills or design processes do you think are the most important to teach? And do you think they're important to teach no matter what country you're in? I'm assuming they're universal. What universal ideas in design do you think are important for everyone to know?

Choi: The most important thing is not a difficult answer. It's not really brain surgery. Understanding and defining a problem. Really simple. That's the root of innovation. Students are expecting me to give a very complicated answer. No, no, no, no.

"OUR REAL CHALLENGE IS ... SPOTTING OPPORTUNITY. WHAT PROBLEMS DO PEOPLE ACTUALLY HAVE? WHAT ARE THE BEHAVIORS? ... THE IMPACTS OF THOSE ISSUES ON THOSE PEOPLE? NOT EVERYBODY CAN DEFINE THOSE THINGS."

Anybody can solve a problem if it's given to them. Our real challenge is the spotting—spotting opportunity. What problems do people actually have? What are the behaviors? What are the variables and boundaries, the impact of those issues on those people? Not everybody can define those things.

That's what they need to learn. They need to know how they can use research tools applied to their design process. What is it? Defining your research question, finding a problem, defining the variables or boundaries and the significance of them. That is design. So another thing, which may be a little bit cheesy, but I'm always telling students, "I don't mind you guys using all your creativity. Actually, you need creativity at some point when you solve the problem. You need an artistic mind during the detail design phase. But, if you design something only for yourself … yeah, that's art. You are in the wrong place."

I have to add one more thing. It's totally okay to start from your own motivation. For example, let's say you saw the moment your grandma had a difficult time gardening because she has arthritis. You want to help your grandma, you want to make the best gardening tool in the world. That's fine. That motivation is great, but you still have to do research to help grandma. And maybe others with arthritis. You have to do research. If you only design something for your self-indulgement, that's art. I'm sorry.

Li: I love this discussion because design and art are so intrinsically linked in some ways.

Design is a beautiful place where art and engineering, creativity and analysis—they can all work together. They can coexist. Do you see differences between how artistic students work and how engineering students work?

Choi: For aesthetic things, everything is possible. In practice, if you look at a lot of beautiful things, you start to level up your aesthetic eye, but fundamentally, the aesthetic aspect is a gift. That doesn't mean you shouldn't level up your aesthetic view. Some engineering students may never want beautiful things. I'm not trying to generalize; some students may never have an aesthetic eye. But my experience is that if engineering students learn design methods and have empathy for people, then the result is always amazing. I do believe that when you're working with engineers, the chances of being successful in the market are higher. There's a lot of other factors, like market opportunity, size, all that stuff. But I do respect the work of engineers because they can build a solution that works as defined, which might have ended up being useless if they hadn't taken users into account.

I've had masters students with engineering backgrounds who are now actually amazing designers—user-experience designers. They're able to be very strong designers because they have a strong background in engineering. Their contribution to the group is showing how something is going to work. That's a strong argument for being multidisciplinary. Of course, artistic designers can also level up by acquiring technical skills.

Li: You've mentioned the skill of building and the skill of making. Why is that distinction so important, do you think?

Choi: Even before a functional model, if a designer doesn't know how the product will be made—what kind of materials, and also the basic manufacturing process, designing for manufacture and assembly—then their design is going to suffer. Eventually, you may find the way but you're going to put a kind of limitation on yourself. Another argument to me is balance. It's still having a voice on both sides. When you know the limitations you can actually have more flexibility. Within this requirement, I can be very creative and clever and design a better way. If you don't have the base knowledge you can never try new things. But if you do, you as a designer can be the one who makes the reasonable, purposeful choice.

Li: And returning to the idea of tradition versus experimentation, how do you look at balancing those things? Between doing design in the way that it's always been done—how it's been taught to you—versus the ways that you change your own design process or experiment with new things?

Choi: Don't throw the baby out with the bathwater. Let me explain. There are basic skills that you need to keep. Let's just say interaction design guidelines. Principles. They have been there forever. Usability. I'm not talking about user experience, I'm talking about usability. There's a Jakob Nielsen book, *Usability Engineering* [1993], that talks about this. It's been there forever and a lot of people spent so much time building those guidelines. So think about usability testing and its basis in ergonomics. Those basic guidelines, they apply for physical interfaces. I still strongly believe that teaching physics—you know, the biomechanical part—is very, very important for designing. Balance that with experiments that you can observe. That is good. We need to teach both, definitely. There can be a moment when some people are going to think that experimentation is art. That's fine. They can use that artistic moment applied to the design with research and quantitative data.

These days, user experience [UX] has kind of swallowed up usability. Unfortunately, in a lot of places, UX means web interface design. I actually wrote a journal paper this evening that discussed that UX as a discipline needs to take usability as its core. There's so much research. UX should build on and follow something like the scientific method to expand to new things in a way that validates and enhances existing knowledge.

Li: So some things we do with user experience change based on tools or technology, but some things will never change because they're based on people. We want to make things easier for them, or less confusing or more aesthetically pleasing. In that sense, people feel happy when they see something beautiful. Design is designing with emotion as well.

Choi: Yes, this reminds me that beautiful things are important. But, think of a designer seeking funding. If a designer is making something that cannot sell in the market, then that's not a good investment. We have to get back to finding the opportunity and the market. Two things I don't need to stress anymore are human-centered design and

beauty. I want my students to have the skills and perspective to be able to beat problems. You can use art, engineering, computing ... whatever. Those are a means to an end. You need a little bit of all of it, these days, to really solve the problem with design, to make life better and get the environment in a better place.

My chair is named after Sir Terence Conran, and his design philosophy is very simple. It's exactly what we just talked about. Your products should function well and efficiently and improve quality of life. His famous quote is: "Good design is ... 98 percent common sense. But what makes the subject so interesting is the other 2 per cent, what you may call aesthetics. Many products are demonstrably good, but those with that extra 2 percent have a magic ingredient that places them in another category altogether—it is the difference between something that is perfectly acceptable and something that is so special that everybody wants to own it."

Li: It's a simple truth, but very hard to practice. Easily said, it's really hard to do well. Simple, easy to use. Beautiful, delightful. Improves the human condition. I started small and I ended very big. It's not easy to do that.

Choi: Simple is more difficult, you know. Always adding things is easy. A diet is more difficult than eating things, right? This is really important, especially with new technologies that are increasingly available, because there always seems to be a tendency to think that just adding on some technology by itself is going to be a solution to a problem because it can appear simple. Just by using it, things will somehow automatically take care of themselves. But a technology is just a tool, not a magic bullet. And like any tool, it's only useful based on how you use it. Using a technology without considering its impact takes our humanity out of design.

The same is true for AI. It can be a tool with a huge amount of potential, but it cannot be a replacement for human elements. What about generative design? Generative-design machines can explore vastly more potential design directions than a human—a thousand, or millions. But that creates another problem. How do you sort out the ideas that are potentially good from the bad out of a million possibilities? Machines are also good at filtering. But just like a program, it takes a human to direct them on what to do. AI abstracts the user away from the problem because you don't always necessarily know how it gave its answers, or even if they're any good. So the challenge, as with any tool, is to figure out how to use it in a way that's both useful and valid for what it's being used for. Those are the kinds of issues that designers will need to be able to explore. The problems and available tools are always changing, so I want my students to be able to make smart and considered decisions about them.

OPPOSITE: Three different contexts, same task. How might region, culture, and infrastructure shape this activity?

4.2 | Expanding Your Frame of View

"Frame of reference" is a concept used in physics and astronomy—an abstract coordinate system with origin, orientation, and scale specified by a set of reference points. Without it, the math calculations driving the behavior of a system could not be derived. In nontechnical terms, this means that your point of view matters. It is a scientist's way of agreeing with ideas covered in the concrete versus abstract art discussion earlier in this book (see pp. 40–41). Expanding your frame of view is examining your product through different frames of reference. Let's take a closer look at what I mean.

Let's assume you're riding the bus, standing up, with a billiard ball in one hand while the other hand is holding onto the railing. You proceed to throw the ball up in the air and then catch it. From your point of view, with the coordinate system set as it is in the illustration at the top of the opposite page, your ball moves in a perfectly vertical up-and-down motion on the scale of about 20 inches (50 cm)—the Y-axis of this coordinate system.

Let's now change our frame of reference and take on the scene from a bystander's perspective. From where this person is standing, the bus is in motion over this time. From the bystander's point of view, with the origin at your left hand, and a new coordinate system relative to them, the ball's motion takes on the shape of a parabola, having a height of 20 inches and a width of about 40 feet (12 m)—depending on the speed of the bus—along the X–Y plane. The motion made by the ball is completely different, seen from their point of view.

We'll make one more shift in frame of reference, still trained on your left hand, but this time from the vantage point of a telecommunications satellite, orbiting at more than 36,000 feet (11 km) above the Earth's surface. The satellite maintains its position above the Earth, moving with the planet's rotation. It is directly above the bus when you first throw the ball so it maintains its position relative to that.

We'll set a new coordinate system centered at your left hand at the moment you threw the ball up. However, due to the distance of the satellite from the Earth's surface, the distance traveled by the ball is so small that it's negligible relative to the speed at which the satellite is traveling, which is the same direction as the Earth's rotation and, in this case, the bus's direction of travel. The ball's motion is a mere speck on the Earth's surface, and the satellite, which is providing satellite radio service to millions of people, wouldn't care about it at all.

Which of these frames of reference is the right one to use? In physics, it would depend on the analysis you were wanting to carry out. If you were trying to understand the ball's complete motion over time, then the bystander's view would be best.

This is the case in product design as well. Designers often mistake the type of problem they face when creating solutions. Naïvely, they only focus on the final form or look of their product, without a greater consideration to context. They don't realize that we live in a social system that has interconnected parts. There may be systems that are relatively simple, such as currency, where the value and use of a particular dollar or euro is equally established everywhere within a region, through to more complex systems. While being complex, some

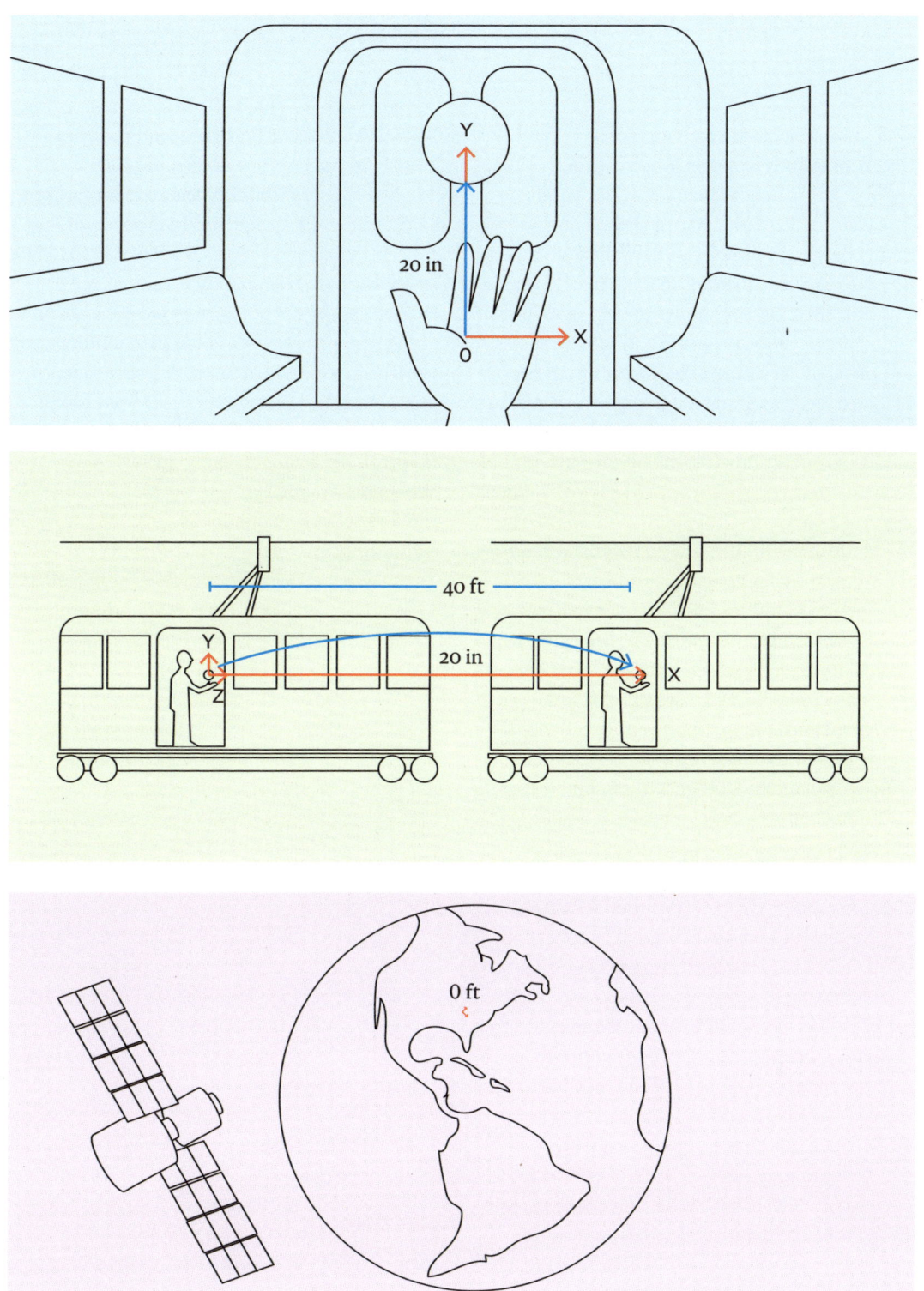

systems can be quite organized, such as the transportation system for goods within a country. There may be many ways for goods to travel (truck, boat, and airplane), but the routes they take, travel duration time, and ports of entry and exit are all known.

On the other hand, there are complex systems that are disorganized—for example, accurate representation within the social-media ecosystem. We may know the handful of social-media channels that most people use, but new social-media avenues are being built every day, and some are targeted at specific groups not in the mainstream, which then produce an echo-chamber effect, creating distortion and reinforcing misinformation. When it comes to representation, there is no single arbiter or universal standard of accuracy, and most of us in the audience lack the means to fact-check references with original sources all over the globe, so we can only trust the intentions and trustworthiness of each individual contributor.

This is a cautionary tale for designers whose products fail in the market because they have failed to identify the correct frame of reference. Misjudging your frame of reference is often a lack of contextual awareness. A toy designer may think, as a result of using their empathetic skills, that a toy should be solely focused on delighting a child. But without taking in the broader context—the person who'll make the toy (the manufacturer), the person who'll sell it (the retailer), the person who'll buy it (the parent), and the environment where it'll be used (the playroom)—they've really missed the complete picture of the product. And the products created by industrial designers are mass-produced, so while you may think the object you created is being enjoyed by someone as a singular experience, that same exerience is actually being replicated by thousands of people all over the world.

Expanding your frame of view, then, means conscientiously adopting other ways with which to look at your product's impact and production. In this chapter, I will introduce several tools that you can use to consider expanding your frame of view, your "cone of vision," which will entail thinking both creatively and critically about your product's context.

"Culture is the widening of the mind and of the spirit."

Jawaharlal Nehru

"WHILE YOU MAY THINK THE OBJECT YOU CREATED IS BEING ENJOYED BY SOMEONE AS A SINGULAR EXPERIENCE, THAT SAME EXPERIENCE IS ACTUALLY BEING REPLICATED BY THOUSANDS OF PEOPLE ALL OVER THE WORLD."

Tool | Powers of 10

The Powers of 10 (or 10x) tool is an analysis and observation technique based on a short film of the same name by the prominent American designers Charles and Ray Eames in 1977.

The Eameses made major contributions to modern architecture and furniture. They also worked in the fields of industrial and graphic design, fine art, and film. They met (and were influential) at the Cranbrook Academy of Art in Michigan. I first saw this documentary film in my undergraduate fine-art days and it made an impression on me then as a wonderfully visual piece of art that also explained a complicated mathematical concept (zooming out—then in—from the starting point of a couple enjoying a picnic by Lake Michigan to explore the relative scale of the Universe based on a factor of ten). In fact, the Eames Office often collaborated with famous scholars in the making of their films. *Powers of Ten*, which can be watched in full on the Eames Office YouTube channel, has also gone on to influence many filmmakers who love employing the single steadycam one-shot.

The tool based on this principle is a method for designers to use when they're unpacking complex issues that play out at multiple scales. Think about zooming out in stages from the view of the activity you're observing right now, as shown in the upper portion of the diagram below. You're imagining all the possible reasons why a problem might have arisen, then thinking of a bigger cause that exists behind those reasons. By thinking of your bug as a symptom, you're in effect conducting a "creative" root-cause thought exercise. In other words, think of the bug you're seeing as a symptom of a much bigger bug. What is the bigger issue at hand?

Applying the tool

Here's a hypothetical case to demonstrate the use of the tool. We'll begin at the center of the model below, choosing an air traveler to represent our observable action (see p. 174). Let's say the activity we just observed was the traveler missing their flight, which has just pulled away from the gate. The traveler is now standing irate in front of the boarding-gate kiosk. We now slowly begin expanding our field of view. We zoom out slightly and question what might have caused this issue. We start at the boarding kiosk. It could've been a personal issue between the traveler

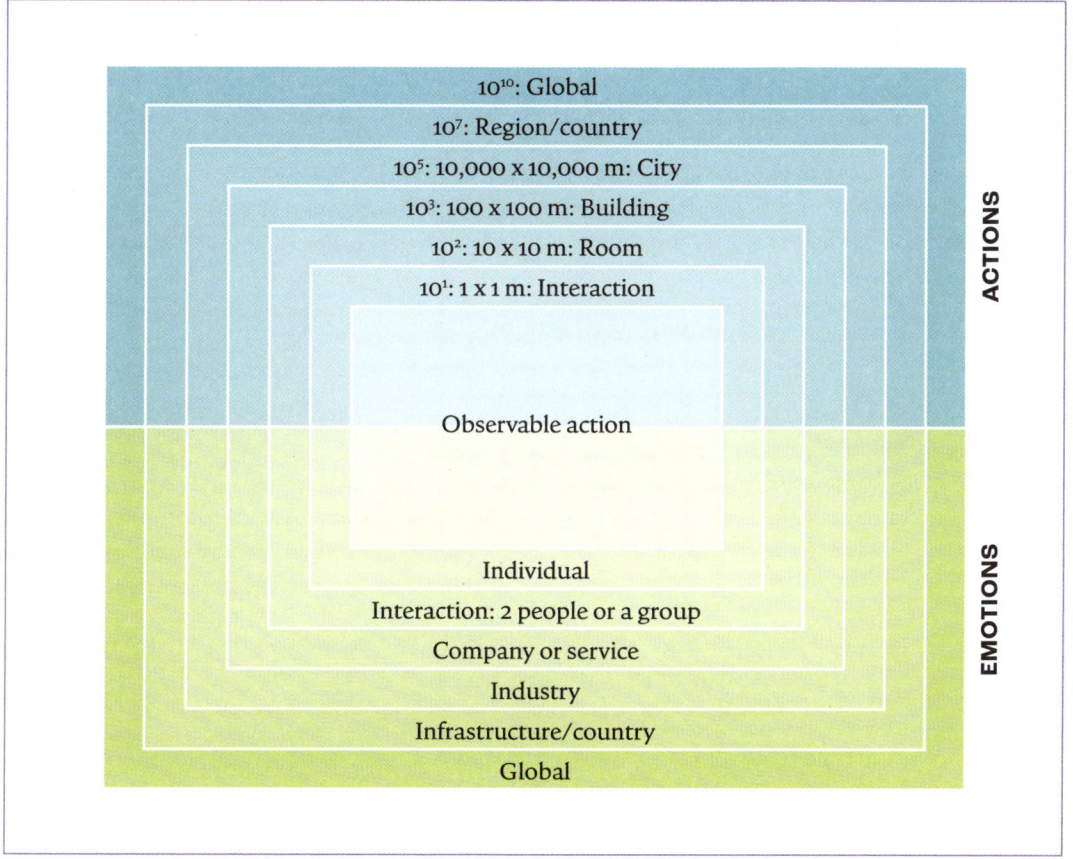

and the gate agent, who decided to uphold the policy of closing the gate doors ten minutes before departure and make an example of the customer, rather than choosing to be lenient and holding the gate open that extra second. The solution would be additional training in customer service for gate agents, right?

Let's zoom out a bit, though. The traveler was only a minute late, so maybe it wasn't down to a decision by the gate agent, but the layout and organization of the airport itself. The security lines were too slow, delaying the traveler, and once through security, the tram to take them to the correct terminal wasn't running regularly. Then the solution would be a more efficient airport layout.

We can zoom out farther still. What about if we consider the placement of the airport relative to the city center. Airports are always located well beyond the outskirts, due to space, noise, and airspace considerations. Traffic along the route to the airport can be an issue—one that consistently causes delays for all travelers, especially during holiday periods.

We can continue to zoom out and see yet more factors that may be in play here. As we do, more societal factors come into play. We can question the role that airlines play within the transportation system, because multiple modes of transportation are needed to coordinate this trip. Trains, cars, and taxis or shared rides all connect to the airport, so the system is interconnected. What role does airline travel play, and how does it contribute to this greater system?

Farther out still we can consider the global transportation system that's used to move people around the world, and how it affects immigration between countries. How does this system cater to the movement of people? What type of people and income level does it cater to?

Finally, on the lower half of the diagram, below the central square, we can complete the expanding square frames with implicit questions about thoughts and feelings. How does the traveler perceive their situation? Use your empathy skills and see if you can discern what the different feelings might be, from their attitudes toward the customer service within a specific airline, and more broadly, how they are treated at airports, through to how they feel about air travel in general, finally expanding to the issue of travel nationwide and internationally, and social stigmas around security or immigration.

Try this yourself using a blank template. Start with a person and an activity or interaction, then use the Powers of 10 to expand (or contract) your frame of view to uncover requirements that you might not have known about on first look. Good luck!

Download the template and then post your Powers of 10 by using the QR codes.

4.3 | Trends, Technology, and Culture

Contextual awareness is one of the design behaviors that can be elusive for young designers. They aspire to create products that are socially impactful and contribute to the zeitgeist, but they're usually not sufficiently cognizant of the cultural forces at work. They just trust that whatever they draw, "someone" will like it. Similarly, there are a great many technologists who, after inventing something in the lab, find themselves so enamored of their new technology—the satisfaction that they've achieved something that previously couldn't be done—that they forget to ask why they've done it, or whether it should've been done in the first place.

There's an implicit bias that any new technology is useful, when that's actually not the case; there are a great many useless patents collecting dust in the world's databases. This is often the situation with failed entrepreneurial activities, where a technology is invented in the lab (just to prove that it could be done) and then the creator tries to find a market for it within society— the "hammer in search of a nail" syndrome. Instead, you must find the human need first, and then select or develop an appropriate form of technology, relevant to human activity.

Being contextually aware means having a sensitivity toward the connections between customer trends, the production landscape, culture, and technology. Collectively, these determine the product's point of view, which in turn determines adoption, effectiveness, and success. While many of the tools presented in this chapter are analytical in nature, they also require abstract thinking, and must be validated with stakeholders.

Humans and the products they make function within societal systems. Sociology teaches us that society can comprise many subcultural groups, or tribes, which all inform the greater culture. Through these cultural associations, the nonmaterial culture is transmitted through stories. Values, beliefs, and morals are all representative of a cultural value system. Further, social anthropologists refer to the mass-produced products we make as "material culture"—artifacts that humans use to survive, define social relationships, represent their identity, or improve their state of being.

As detailed in the section on entreprenuerial sustainability (see pp. 102–5), the take-up of a new technology also occurs through an adoption curve. It is accepted by the market in waves, starting with early adopters and technology evanglists who are less risk-averse and willing to take a gamble on something new (that may or may not work), then moving into the mainstream, more risk-averse, conservative markets. It is the adoption waves that create trends and fads in society. As the dialogue between technology evangelists helps to convince the more mainstream or conservative among us of the benefits of utilizing the new product, the cultural associations of that product change—it becomes accepted on a wider scale, and a larger demand is created as a result.

The objects that we design are made of parts that are, in turn, designed by others. In mass production, this relationship brings additional complexity to the design task. For example, an automobile may be designed by an automotive manufacturer, like Volkswagen. However, Volkswagen do not design the tires for their

OPPOSITE: A banana farm in Ecuador.

vehicles. Those come from another company and factory owned by Continental Tire. Or Pirelli, or Goodyear. Volkswagen may procure parts from myriad suppliers. Instrument dials, light bulbs, screws, and computer chips are all sourced from different companies, and perhaps assembled into a group (called a subassembly, an example being an interior instrument panel), which then ships to Volkswagen's central location for final assembly as a complete vehicle. The parts we select for our products are all part of a large, logistical supply-chain system, and unless we're sourcing locally, part of a global economic system also. Parts, production, technology, and culture are all interconnected. Let's take a hypothetical, humanitarian design example to illustrate some of these relationships.

Consider the development of an energy system for a rural farmer in Ecuador, where many farms face rationed electricity. Without context, a solar array might be the first thing you'd think of to solve the problem. In agriculture, the use of solar arrays for farming is commonly called "agrivoltaics"—employed by many modern farms to help generate power for equipment, irrigation systems, and living quarters. A technology like agrivoltaics might work in the United States or France, but would it make sense to use it in South America? Let's consider the needs of the farmer first, whose primary crop might be bananas.

In the foothills of the Andes mountains, bananas might grow very well, but the terrain may make a solar-panel system difficult to import, install,

and maintain, especially if it had to be put on stilts or a roof structure high above the crops. Solar panels lose efficiency when they get dirty or dusty too, so keeping them clean would be another issue, as there'd be no workers trained in the maintenance of these types of structure. Further, the solar panels themselves would need to be shipped. In the States and Europe, solar panels are manufactured locally so are more readily available, but in South America they'd have to be imported. If an electronic component, such as an inverter or an electric motor used to position the panels, were to break, that part would also have to be imported. Then someone trained in the system would have to perform the installation and repair. Because no one would be available to carry out maintainence and repair, the system would end up languishing in the fields. You'd also need to consider the costs of distributing this type of system to multiple farms in the area. For all these reasons, what might have seemed like a good idea fails within the real-life context.

What might be a better solution? It would take a better understanding of the farming environment and the infrastructural challenges of the country to make a system that would help farmers better. Whether it's a system of aqueducts, connecting each farm with rainfall that then powers a hydroelectric power station, or a collective wind farm shared across the area that every farmer pays into, the onus is on the designer to understand the connections between way of life, manufacturing capability, and adoption for success.

BELOW: Agrivoltaics being used on a farm in France.

PESTLE

Devised by Francis Aguilar at the Harvard Business School, this is a very helpful technique to use when thinking about the interplay of technology and culture for the business and/or culture that you're investigating. Created to help businesses identify the issues that might help or hinder market adoption, the PESTLE framework groups relevant factors in six different categories, establishing a detailed overview of the context.

P
POLITICAL

Political factors include: policy frameworks, labor-market policies, government policies, competition oversight, political stability, security requirements, and subsidies.

E
ECONOMICAL

Economic factors include: economic growth, inflation, interest rates, exchange rates, taxation, unemployment, income, business cycles, world trade, and availability of resources.

S
SOCIAL / SOCIOLOGICAL

Social factors include: values, attitudes, lifestyles, consumer trends, demographics, income distribution, education, population development, security, as well as aspects within society, such as family, friends, colleagues, neighbors, and the media.

T
TECHNOLOGICAL

Technological factors include: new technologies, technology effects, research, development speed, new products and processes, product lifecycles, technology investments, and government research expenditure.

L
LEGAL

These factors relate to the legal environment in which the business operates, and include: legislation, tax policies, trade barriers, and tariffs.

E
ENVIRONMENTAL

Environmental factors include: material resources, disposal, emission regulations, energy, transportation routes, life cycles, effects of the ozone hole and global warming.

4.4 | Organizing Human Needs

Contextual awareness hinges upon the interplay between human needs at the individual activity level, with contextual requirements of the activity within the environment, and extending further through to cultural or tribal norms at the societal level. Because the products we make function within the systems of a society, we also need tools to help us relate to those diverse needs, in order to "cut cubes out of the fog."

We've already looked at Maslow's hierarchy of needs as one way to organize human needs (see pp. 128–29). We've also seen that technological and production concerns are culturally and geographically dependent. With the interplay of people's individual wants, their placement in a greater culture's requirements, and technical constraints (material logistics, manufacture, supply chain), we need to organize needs to help manage some of these complexities.

We can all recognize that some needs function at the individual level, while others function at a cultural level. We also know that there's a difference between the need to breathe, the need to believe in a higher power, and the need to own the latest iPhone. We therefore have to organize human needs and abstract them. Expanding your view means moving from a concrete need that's readily observable and asking which more abstract needs or other influences may have caused the observable need. Without identifying the differences between needs, as well as what motivates them, we're not able to channel our creative mindsets into something productive. Human needs at a cultural level are more abstract, implied, and unspoken, but they also involve a greater number of people. Organizing those needs requires us to tap into different mindsets—employing our analytical ability to identify which needs may surface and drive others, and our creative and empathetic skills to identify implicit, hidden needs. The latter often go unspoken, and our ability to give them voice in order to create products that address them is a secret superpower.

Organizing human needs begins with observing and identifying as many needs as possible from the stakeholders involved with your project. From your participation strategy (pp. 74–75), you will have identified key people to observe or interview. From your field notes, you should be able to mind map many potential needs and get them down on Post-it notes (one need per note). These can be used to uncover any hidden relationships among the needs. You can, of course, also utilize any of the other tools we've introduced so far to aid this process. Empathy maps (pp. 126–27) can help to identify the needs observed in an activity or scene. AEIOU notes (pp. 88–89) of an observation that involved your stakeholders working through different issues can help you to empathize with them and posit potential needs (using verbs). You can also use positionality (pp. 132–33) to encourage your cognitive empathy, method-acting techniques to deepen your emotional understanding (pp. 140–43), or the open-ended, semi-structured interview technique (pp. 152–56) to extract stories from your stakeholders. Through these empathetic tools alone you will likely identify a myriad of needs.

Now, explore your frame of view through a Powers of 10 exercise (pp. 172–76). Or create a journey map (pp. 156–57) for each stakeholder, outlining their participation in the process or activity you've identified. Whichever methods you choose, once done, put the results up on a pinboard/magnetic whiteboard. This is your design empathy research, and your organization of these needs is your synthesis of the situation.

When we organize needs, we look for those that are at odds with each other. What do you think caused this? A source of contradiction is often a golden opportunity for design. Listen for the stories of contradiction, normative behavior, success, and failure, which often hold the values and beliefs that run counter to the accepted order. Design is finding a way to mediate between the two. If you're having a hard time finding insights and connections within your needs, don't worry—I'll be providing two more tools that can help with this task.

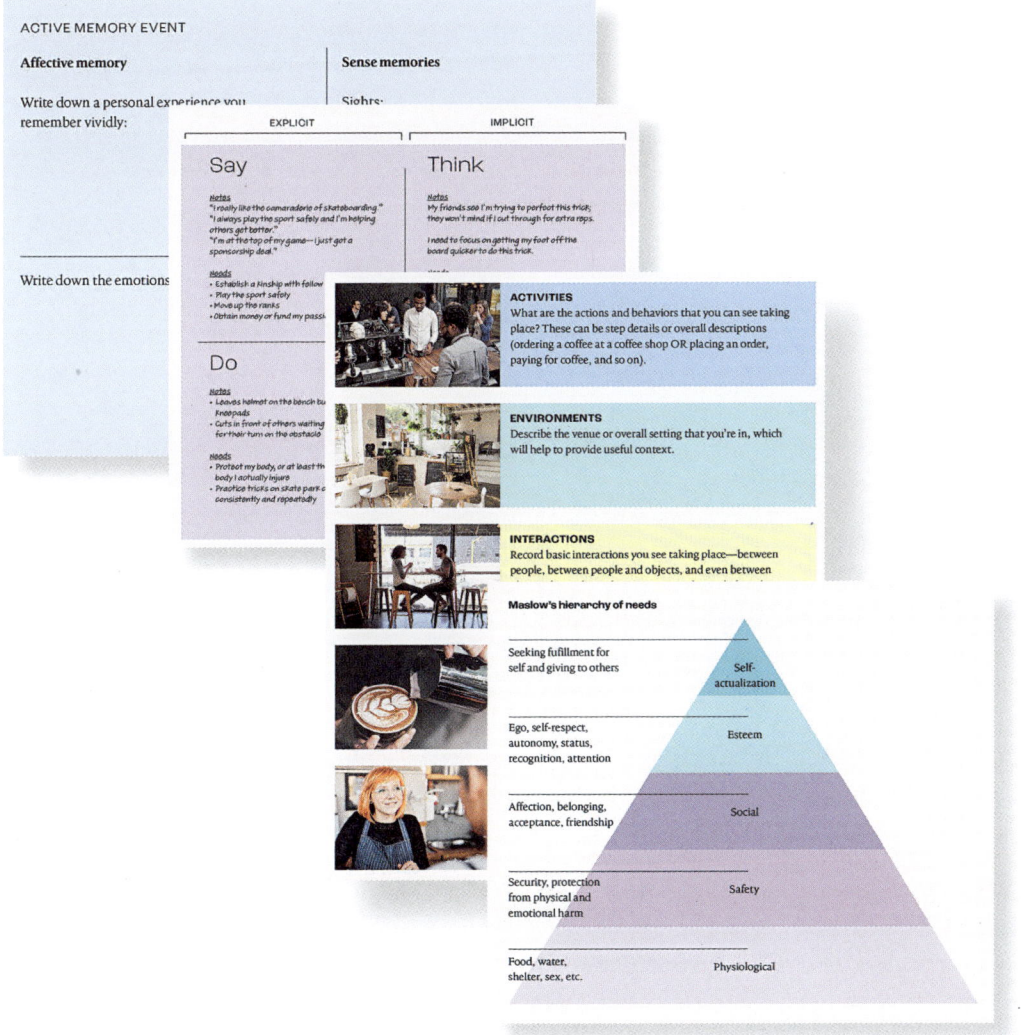

Tool | Common to Qualifier

This tool has its basis in the "total quality management" techniques inspired by successful production systems in Japan, the Toyota Production System being one example. Similar to the well-known technique of asking "five whys" to get to a root cause, the Common to Qualifier tool utilizes a "why–how" laddering technique with a creative spin on identifying human needs. This method was created by Dev Patnaik and Michael Barry at Stanford.

Design solutions have a hierarchy of needs similar to the hierarchy we've explored in relation to humans. For example, the need to drive in a nail is both a human need when building a treehouse and also the function of a hammer. In a why–how laddering technique, you ask why something exists in order to ascertain what might have caused it, and to discover more abstract causes behind what's concrete. Asking "why" something exists moves your thinking up the ladder to higher (more abstract) rungs. To come down the ladder, you ask "how" you might accomplish a goal or fix a problem at each rung. This thought exercise instigates a creative output of potential solutions for each need.

The Common to Qualifier method starts with a very specific need—a "qualifier need," as shown in the sample opposite. We start with the lowest-level need, which has a specific context: a particular activity, and people who want to perform that activity in the same way. Let's ask our first why: "Why does someone want to drink coffee (in this specific way)?" Please note that this a creative exercise; unlike root-cause analysis, there may be multiple needs at each level. This is also an exercise in empathetic thinking, so get into character (as your stakeholder), imagine their potential "activity needs," and pick one for the next level on your chart. An activity need—people in the same context who want to do the same thing— might be "going on a date" (although other plausible activity needs for drinking a coffee might be "completing a homework assignment" or "feeding a caffeine addiction").

Asking a second why ("Why do we go out on dates?") will lead us to a "context need," shared by others of the same age, profession, or region. This might be the need "to get married." The third and final why brings us to a "common need," experienced by almost everyone, such as the need "to be loved."

Coming down the ladder, we can ask "How might we accomplish these things?" Again, this is a creative exercise, so there can be multiple solutions to address each need. For example, at the common level, the need to feel and have love returned is a global system of social connection or alternately faith (organized religion). The need to get married involves a business model, or "families of offerings" (see opposite). "How" you might go on a date encompasses many products and services; and finally, how you might take your coffee would be the features of a coffee drink.

Using the Common to Qualifier method is a great way to both expand your market and make sure you're not working on too narrow an idea. Qualifier needs are usually too specific; they don't allow for design flexibility. If you're only concentrating on existing product features, your design work will be incremental and unlikely to be innovative enough to move the needle. At the other end of the scale, common needs are too universal to allow for specific design solutions; it's difficult to design a system so adaptable that it suits a global market. If you could create a single solution, it would be a powerful design, but for an entrepreneur just starting out, it would be hard to scale to that volume of customers quickly. The strongest design opportunities and most innovative solutions lie in the context and activity layers, where families of offerings, products, and services can address the appropriate level of need. Running through this method, using an empathetic mindset to identify needs at each level, allows you to check your understanding of a specific community's context and activities. A great way to make a specific qualifier need obsolete is to look at the need levels directly above it. By bearing the activity and context levels in mind, you can redefine the needs at the qualifier level.

Common to Qualifier sample

"HOW" COMES DOWN

Common	Need to be loved	Needs of nearly everyone	◀ SYSTEMS
Context	Need to get married	Needs of people of the same age, profession, region, etc.	◀ FAMILIES OF OFFERINGS
Activity	Need to go on a date	Needs of people in the same context who want to do the same thing	◀ PRODUCTS & SERVICES
Qualifier	Need to drink a double decaf skim mocha	Needs of people in the same context who want to do the same thing, the same way	◀ FEATURES

START HERE — "WHY" GOES UP

→ EXIT

Download the template and then post your Common to Qualifier by using the QR codes.

TOOL: COMMON TO QUALIFIER | 183

Tool | Affinity Mapping and 2x2 Matrices

Affinity mapping and 2x2 matrices are tools that go hand in hand to draw out relationships between different needs.

Affinity mapping

This method is used to organize a collection of items into meaningful groupings. It offers a way to organize needs into themes that can be addressed through varying criteria within your design. For example, how might the individual considerations of cost, production volume, durability, ease of use, and aesthetics be brought forward in the design? Knowing the priorities of each criterion can help you come up with ideas. Affinity mapping can provide informal constraints to your process as well as jumping-off points for inspiration.

You begin with a large collection of needs, each written on a separate card. One type of affinity map is called a "card sort." You approach this in one of two ways: "top down" or "bottom up." For the top-down approach, begin by identifying the categories of needs that your design team is looking for. For example, if you're concerned about how affordable a product needs to be, one pile of cards could be assembled under the heading "needs associated with costs." Another category might be usability—how easy something is to use. This technique can also be conducted with stakeholders, providing them with cards that you've already populated with needs in order to pick out their preferences and priorities. You can also provide them with blank cards to help generate needs that you've missed.

The bottom-up, or "open sort," method conducts this exercise, usually with a customer, by listing all their needs first and, from the data itself, grouping the needs into the categories that present themselves. This helps to identify any categories that you might have missed but that are inherent in the data. Outliers—needs the customer had but weren't previously identified top-down—are something to make note of. By sorting the cards into piles of interest, you'll discover which needs fit which dimension of interest.

Bottom-up card sorting

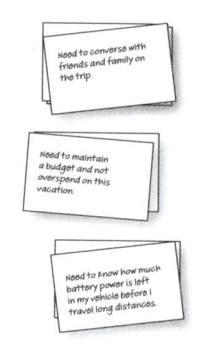

1. Record ideas on cards.

2. Sort them into groups.

3. Label the groups.

2x2 matrices

A 2x2 matrix (two axes, forming two rows and two columns) will draw out relationships between different factors, explaining the "whys" behind them, which can be helpful for deduction or prediction in design. Each axis is best summarized as a specific factor and the lack, or direct opposite, of that factor.

Start by brainstorming potential factors. Go for quantity and keep the axes simple; avoid multiple labels on the same axis. Next, organize the needs by ranking them on a numerical scale. This is better than trying to compare needs against each other. By the time you're in the middle of the data, it'll be too hard to remember which needs were compared to which, so a consistent numerical ranking with a clear definition of what is a 1, 5, or 10 (out of 10) will be easier to maintain. Assuming you have a large collection of needs, you'll begin to see them sorted into patterns based on the chosen axes. Think about what the extreme corners of your matrix might be. Look out for linear relationships, which show a connection between your two characteristics. Pay attention to open spaces. These might indicate a design opportunity, or a keep-out zone (a strong reason why no one has populated them before). Look too for where only one or two needs populate an area—outliers—and question why these exist. As with affinity mapping, a 2x2 can help you determine a strong area or opportunity for design and this direction can be used as a strategic mission statement for your next creative phase, which is idea generation.

Download the template and then post your matrix by using the QR codes.

2x2 matrix

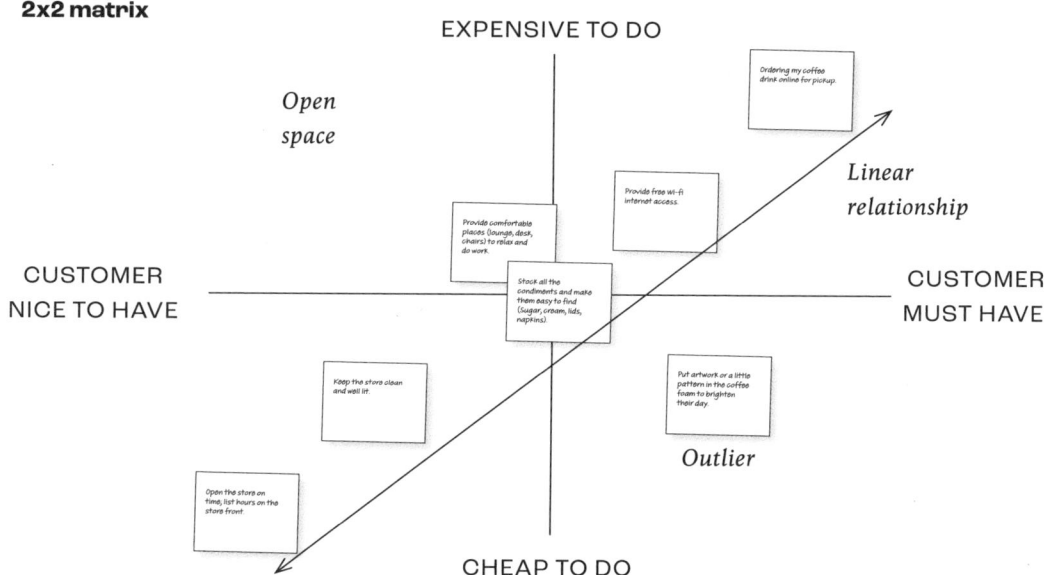

TOOL: AFFINITY MAPPING AND 2X2 MATRICES

4.5 | Timelines and Mapping

When it comes to the context of a product, we've covered expanding your viewpoint, considering the relationship between technology and culture, and organizing needs based on levels of abstraction, from individual to group or culture. However, there are two factors that we haven't yet explored: time and space.

Timelines

Needs in society have longer time windows than technology does, so by considering objects, data, or criteria over a period of time, we're able to draw attention to changes in the characteristics that drive human needs. Timelines are a great way to identify how various characteristics and humanity intersect. For example, in modern life, recording music, saving work documents, and storing movies and photos all revolve around saving digital data. These needs have remained unchanged since the inception of computers, but what has changed are the technologies used to address them. The timeline below shows a range of recording and storage options— from magnetic and optical disks through to solid-state memory and cloud storage. This chronological arrangement enables us to compare and contrast characteristics over time, and to identify the relationships between different needs.

When a new technology comes out, it may at first be more expensive, but in general, the trend

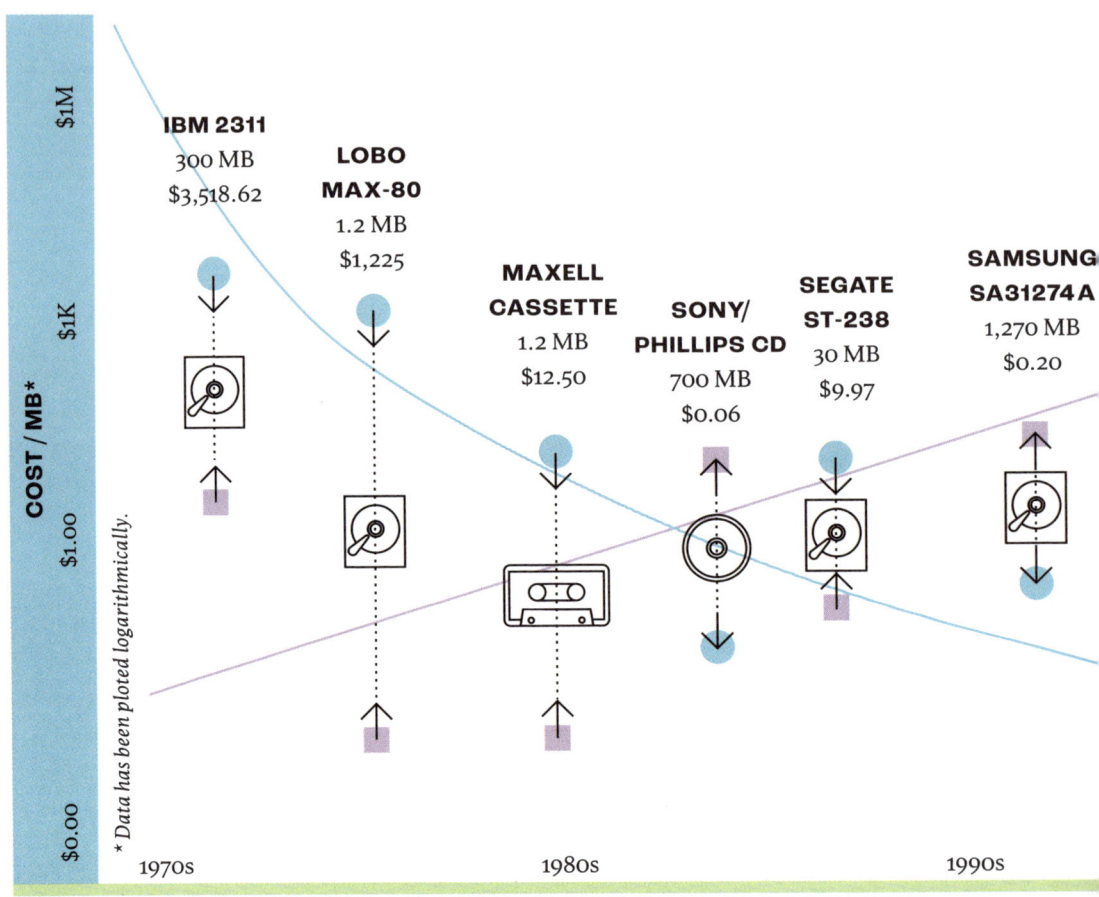

has been steady increases in amounts of storage for exponentially decreasing costs. When the data is plotted, it's interesting that one curve is more linear and the other exponential. What this means is that though storage amounts increase steadily, prices must be reduced in a more pronounced way to affect hearts and minds.

Note too that new forms of storage may appear to cater to particular needs. For example, the cassette tape and CD came about due to a need for portability, versus the hard drives stored in a computer. In the case of CDs, the disk format could only hold music files and could not be erased or rewritten (until CD-RW and CD burners came out). Finally, note that the last data point—a cloud storage service—would be more expensive than the disk drives because it follows a subscription model; with this option, you're renting someone else's disk drives. What determines the price is the perceived increase in value that the cloud service brings. Portability is no longer an issue, and the storage is secure and maintained by someone else. The risk of a failed hard disk, combined with the ability to access your data from any device with an internet connection, is what would entice a customer to subscribe.

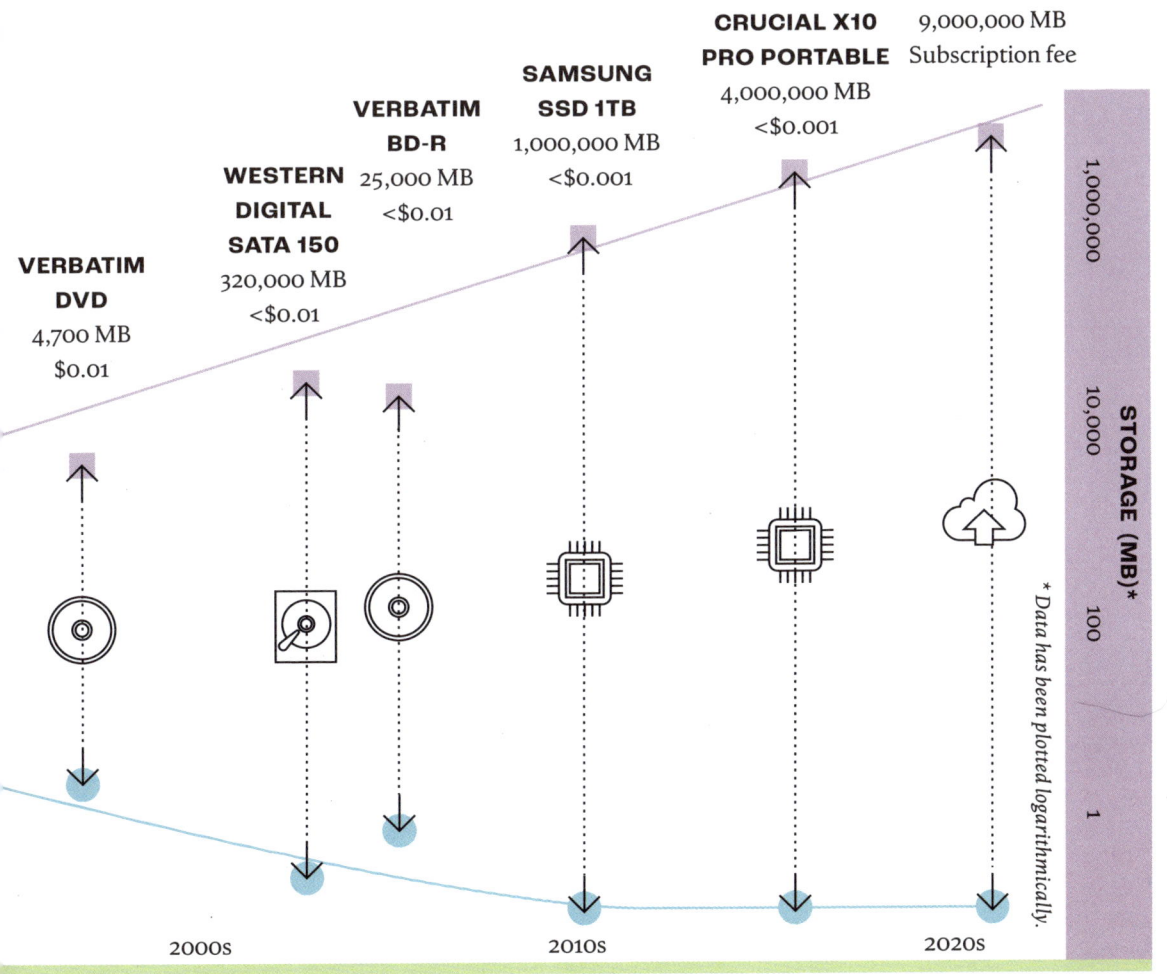

Mapping

Including physical mapping in your toolbox to illustrate needs or criteria based on location helps to give your work cultural comparison and context. We've already covered affinity mapping (see p. 184). A different and famous early example is shown here. This is English physician John Snow's dot map of a cholera outbreak that occurred in Broad (now Broadwick) Street, in London's Soho district in 1854.

At the time, the transmission of germs was not well understood, and cholera was thought to be caused by a miasma—"noxious vapors" in the air. Snow, however, was skeptical about the theory, so he investigated the deaths that had occurred by location. Mapping these as bar lines on a street plan revealed a cluster of cases in a particular area, providing a pattern that helped pinpoint the cause of the outbreak to a specific water pump on Broad Street. By using the map as evidence, he was able to convince the local council that the disease was waterborne. The well pump was disabled and the outbreak ended. By using data visualization to uncover a relationship between the individual cases, Snow's map changed our understanding of microbes and the spread of disease.

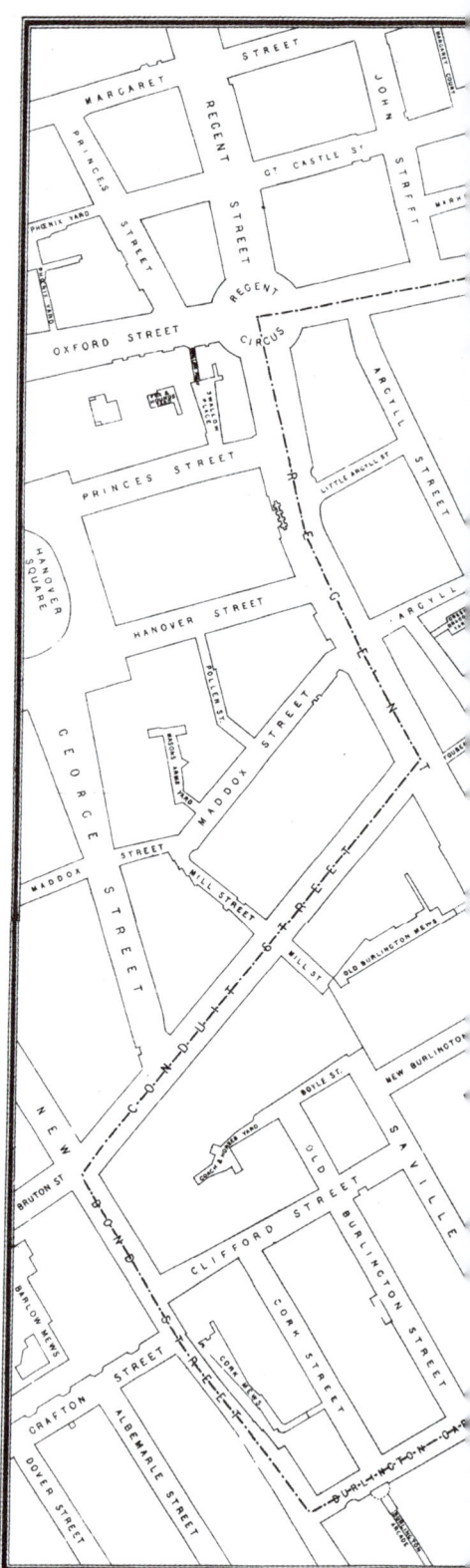

RIGHT: Dr. Snow's map of the Soho cholera outbreak of 1854. By ascertaining and recording the number of cases (black bar lines) by location, he was able to pinpoint the water source that had caused the outbreak.

Tool | 9 Windows

The 9 Windows method combines a few of the techniques we've covered already. It was based on the "theory of inventive problem solving," or TRIZ (an acronym for the Russian *teorija reshenija izobretatelskikh zadach*), created in the mid-20th century by Genrich Altshuller, a Soviet inventor and science-fiction writer.

TRIZ is founded on two basic principles. First, that somebody, sometime in the past, has already solved your problem or one similar to it. The creativity lies in finding that solution and adapting it to your current situation. Second, as we've mentioned before, that design opportunities are identified by resolving the contradiction between competing design objectives or factors.

The 9 Windows method

TRIZ uses a Powers of 10 approach (see pp. 172–75) to both expand and contract a point of view, then combines it with a timeline to identify contradictions and opportunities for solutions. The basic approach to using this tool is to start with your current problem and place it in the center of what will form a 3x3 matrix to define the issue in a clear and simple way. Then, from the center, you expand up and down. The top horizontal level of the matrix is the super-system (or macrosystem). This is the external environment and components that the problem interacts with. The sub-system (or microsystem) on the bottom level consists of components or parts of the problem.

Next, you can approach the problem by viewing it as a system, monitoring all the interactions and the stakeholders that may exist while zooming in and out of that system, both in the past and into the future. Your timeframe steps should be consistently forward and backward. Determine the most powerful principles applicable for the respective problem and come up with innovative ideas for improving the strategy.

Opposite is an example of the 9 Windows method used for the design of a toaster. The issue at hand is resolving the toasting of bread, or preparing food, quickly and compactly. You can see that we started with the present-day toaster in the center square, then looked at the environment of the product, as well as the components and technology related to its use. Then we moved backward and forward in time for each of the other two sets. By looking at how interactions have changed from yesterday to today, and between super and sub-system, you can theorize potential avenues to advance your product design.

OPPOSITE: The 9 Windows method is used to predict and anticipate what a product might become in the future, based on its past and existing environment and components.

Download the template and then post your 9 Windows by using the QR codes.

190 | CHAPTER 4: CONTEXTUAL AWARENESS

9 Windows sample

	PAST	PRESENT	FUTURE
SUPER-SYSTEM			
SYSTEM			
SUB-SYSTEM	**Material:** Chromed steel **Heat source:** Copper heating element	**Materials:** Stamped metal and plastic **Heat source:** Nichrome wire	**Materials:** Bio-based plastics **Heat source:** Induction plate

TOOL: 9 WINDOWS

Tool | Era Analysis—Material Culture

This tool combines the use of typologies (the study of the sorting of types according to characteristics) and material culture (study of artifacts). We compare the same items over time and draw inferences based on their presentation.

With this tool, surprising relationships can emerge that signal the zeitgeist. Let's explore it by looking at the evolution of Batman.

Created by Robert Kane and Bill Finger, Batman first appeared in *Detective Comics* No. 27 in 1939 as a crime-fighter hero intended to appeal to teenage boys. Early comics depicted him as a detective solving crimes with the aid of an array of crime-solving accoutrements, including a streamlined costume, utility belt, vehicle (batmobile), and a boomerang-like weapon (batarang).

Batman over the years

1960s

1980s

CHAPTER 4: CONTEXTUAL AWARENESS

Let's look at a few of his depictions over the years, focusing on television and movies.

COSTUME

Batman's costume has certainly changed through the decades. The colorful spandex bodysuits worn by Adam West and Burt Ward (as Robin) in the mid-1960s television series were used for dancing as much as fighting criminals, with each punch landing with a giant "Pow!" across the screen. Batman during this era was presented as law-abiding, honest, and upstanding. Fast forward to 1989, with Michael Keaton starring in the first of the Warner Brothers series of movies, and the color has been replaced with a rubber, foam, and latex suit with Herculean muscles molded in (down to the pectoral nipples in later

2000s

2010s

movies of the era). This is a decade of excess, from big hair to large shoulderpads for women's suits, and Batman's costume fits right in. However, his excessive muscularity takes a turn in the 2000s. Christian Bale's outfit goes from fake muscles to a very real "Nomex survival suit for advanced infantry. Kevlar bi-weave, reinforced joints," as only Morgan Freeman could have described it. By now the costume is a military protective suit appropriate for the special forces. When Batman fights Superman in the 2010s, Ben Affleck's costume moves from an armored vest to an "Iron Man" styled suit of armor—a robotic exoskeleton to take on the alien representing truth, justice, and the American way. In the 2020s, Robert Pattinson's crime noir character is a younger representation, and his costume is more of a leather and metal-reinforced "biker jacket," albeit bulletproof, for a vigilante who's just starting his crime-fighting career.

BATMOBILE
The vehicle that Batman uses, like a trusty horse, is a character in its own right. The first Batmobile, seen in the comics of the 1940s, is a cross between a regular vehicle, a Mercury Eight coupe, and a large bat-shaped hood ornament, resembling a face (which doubles as a battering ram). It has a bat-inspired tail fin. This motif continues through to the 1960s vehicle, a Ford Futura chassis with a bat-faced front end, large tail fins, airplane-canopy windows, and jet-engine propulsion customized by designer George Barris. In the 1980s, the vehicle becomes even more operatic, incorporating the face as a cowled cockpit that opens like a fighter jet, and includes both jet engine and machine guns.

It transitions in the 2000s to a "Tumbler," which is a cross between two military vehicles, a stealth jet fighter (F-117 Nighthawk) and a military off-road vehicle (Humvee, or HMMWV). The 2010 version is a blend of the two prior vehicles, and the 2022 incarnation goes back to earlier roots, with a rugged muscle car as the basis for the vehicle, with a battering ram type of front end and jet-engined afterburner in the rear.

BAT INSIGNIA
The logo, or bat insignia, takes its form in a variety of ways as well. Used as a typographic treatment in the 1960s, the insignia is a yellow piece of fabric, but it turns into an embossed gold and black emblem iconic with 1980s bling. Christian Bale's 2005 chest emblem is spray-painted black for stealth but it inherits a function: that of a Japanese *shuriken*—a ninja star, or throwing star. That weapon turns a bit darker in director Zack Snyder's 2016 version, turning into a heated brand for permanently branding the skin of Batman's enemies. The weapon motif carries on to the 2022 Batman, but instead of a weapon to harm evildoers, it's the actual weapon (pistol) that was used to murder Batman's parents, filetted, splayed open, and displayed on his chest. It's not as deadly as a throwing star or a branding iron, but no less dark.

INSIGHTS AND TAKEAWAYS
What insights can we learn using this technique? We're able to see shifts in cultural attitudes by era or decade. Remember, the target audience (adolescent boys or young men) has not changed (all the movies are PG-13 or below), nor has the core concept, Batman himself. However, the depiction of

our hero reveals some surprising insights. Whereas evil is fought with dancing, talking, and punches that never land in the 1960s, the escalation and acceptance of violence as time progresses—both in entertainment and as justification to solve issues—is noteworthy, down to the very graphic, realistic depiction of violence in contemporary movies. A very interesting turn occurs in the 2000s, which can be traced back to 9/11 and the "Wars on Terror" in Iraq and Afghanistan. In a day, the US's innocence was taken away, and evil can now only be fought through overwhelming military means. Further, cynicism with our established institutions, like representative government, has escalated, and erosion of trust in established law enforcement can be seen from iteration to iteration. Batman has

Download the template and then post your era analysis by using the QR codes.

charted the rise of the anti-hero: operating above the law and violating the rights of all others is the domain of Batman in his Dark Knight guise. He's certainly not the white knight in shining armor we associate with heroes. This trend increases so much that Batman and Superman are enemies by the 2010s. Where will Batman go next?

BELOW: Luckily, and perhaps being more than a little self-aware, DC Comics and Warner Bros. released *The Lego Batman Movie* in 2017 to help lift our spirits out of the rage and darkness.

4.6 | Design and Business Strategy

We mentioned previously that novice designers usually "jump to solve" and are quick to make several simplifying assumptions in order to arrive at a product solution without further analysis or abstraction. They see someone limping and immediately design a crutch, without using empathy or contextual awareness to establish how the person developed the limp in the first place, yet the myriad potential causes for the limp will affect the resonance of the design.

Designers who do employ these mindsets are not only taking the time to trigger their compassionate concern, they're also attempting to abstract and organize human needs, and bring contextual analysis to the problem. How do cultural forces, trends, technology, and societal concerns shape the project? Practicing these behaviors contributes to successful products, and are the hallmarks of a more sophisticated designer.

In this last chapter, we've covered the last phase that takes place before moving on to ideating products. Where empathy and contextual awareness have been used to observe and abstract human needs, we must now be creative about harnessing our ideas and compassion and formulating a cohesive business strategy. The role of "design strategist" is usually given to a designer for exactly this purpose—that is, understanding the customer's needs in context, and relating that to the company's brand, mission statement, or core purpose. All these factors combined determine what product to make, whether it be a single item, a collection, a new brand, or a whole new subsidiary business. Synonymous titles include "product planner" or "brand manager," which may also be a role in the marketing department of many companies.

In moving from an analysis of human needs and context, we now shift into a creative mindset to persuade others to act. Rather than being a solo practitioner in a single studio, in reality you'll have to advocate for your product—persuade others of the product's ability to contribute to the bottom line, both financially (for profit, if a commercial entity) and through social impact (for humanitarian causes).

How do we persuade people and compel them to action? Ultimately, financiers or production have to green-light a strategy first, in order to allocate the budget for you to create the physical products, which are the tactics used to execute the strategy. Our task is to provide specific and actionable directions for how we seek to change the world, and more specifically, what we intend to design. Further, if we can identify the frame of meaning that we'd like to change, we can focus on strategies that affect that change. In this way, we're materializing culture, giving form to the change in meaning that we seek to make.

Design strategies

The following are some ideas and advice for creating design strategies.

1. IMPERATIVE STATEMENT
Most strategies are about persuasion and convincing someone to act, so feel free to phrase your strategy as an imperative statement, a call to action: "Everything we do will seek to lift up those in the most need."

2. FRAME AND REFRAME
Through your empathy skills and contextual awareness, you've probably been able to identify a frame of meaning composed of a contradiction

that's causing cognitive dissonance—and that is the frame you want to reimagine. Identify and think about the frame that you'd like to change. Reframe the contradiction into a future frame of meaning that describes an ideal outcome.

3. METAPHORS

Making a simple declarative statement that helps compare and define two different concepts is an effective way to rally a design team. For example, "Relationships are a journey, and our service provides you with the most intimate tour guide you could have." You can use both current and future metaphors, which state how things are today and how an imagined future tracks from your timeline analysis. You can use the 9 Windows method (see pp. 190–91) to trigger your creative, episodic memory to imagine a better future and think of a simple concept that encapsulates the qualities you want people to remember.

4. "HOW MIGHT WE" STATEMENTS

A great way to start generating ideas is to use a "how might we" (HMW) statement, which acts as a design prompt. HMWs present questions that get to the heart of the matter—the human needs you want to address. When creating HMWs, as with context and activity needs, you need to avoid statements that are too narrow ("How might we make it easier for someone to turn this doorknob?") or too broad ("How might we solve world peace?"). Instead, you want to find the level at which your company can affect change: "How might we encourage more physical activity for seniors?" It can also be worth looking at increasing benefits, removing pain, exploring the opposite, identifying untapped resources, or changing the status quo.

5. ROADMAP YOUR PRODUCT SOLUTIONS

Part of entrepreneurial sustainability is producing at the appropriate level for your company's maturity. A young company may not have the resources to release a mass-produced product globally; they don't have the partner network in sales or distribution yet up to that task. So a more limited product release needs to target the right audience size in the right locale. Profits from the alpha release will then fund a larger beta offering down the line. Future markets as well as further release of new products (as a collection) can be staged as production level increases.

There are a lot of resources for learning more about design strategy and marketing, and entrepreneurial sustainability (see pp. 102–5)—a critical design behavior—can be developed through collaboration with marketers, advertisers, salespeople, the supply chain, and logistics experts in your industry.

"ANYTHING YOU WANT TO BE SON, YOU'LL BE. NO ONE WILL EVER STOP YOU."

Mr. Sanderson, *The Sound of Summer Running* (Ray Bradbury)

From the Author

Congratulations—you've made it to the end. To quote Obi-Wan Kenobi, a mentor from another time and a galaxy far far away, "That's good. You've taken your first step into a larger world." I'm glad you're still here and enthusiastic about embarking on a journey in design.

I've mentioned already that design is not a discipline, but a lifestyle. You live it every day and your experiences contribute to your view of it, which in turn shapes your competency. As a design educator, I often get asked for advice on how to improve as a designer, and it's always a tricky question to answer without context. Are you trying to improve your skills when it comes to visualization, creativity, craft, or empathy, or are you trying to commercialize a product or launch a business? What's your purpose?

"Design" is one of the few words that can be a noun, verb, and adjective all at once. So the sentence, "I'm designing a design that will design a design that's designed" actually makes sense! Ultimately, your journey through life is in itself a design. To quote another mentor closer to home—the Cranbrook-trained designer/artist Matt Kahn—"You're a designer, aren't you? So why aren't you using design to design?" What's the challenge ahead? Who are the stakeholders? Have you empathized with them? Defined how to contribute and improve their way of life? Created options and tested them? If you think deeply about your design process and remain honest, it'll not only guide you in your studio projects but improve your life too. Being restricted to your own frame of reference will prevent you from achieving your ultimate potential on the planet. When you wield design well, however, you realize that you're not helpless, or at the mercy of a large wave—you're commanding your life through uncharted waters with a host of fellow navigators.

For those seeking less esoteric advice, I've summarized three general pointers on the following pages, which should prove handy as you progress. And in the meantime, thanks for your contribution to a community of like-minded creative practitioners. I wish you the best of luck in all your endeavors.

Wayne Li
Oliver Professor of Design and Engineering,
Georgia Institute of Technology

OPPOSITE: In 1967, Kathrine Switzer became the first woman to officially complete the Boston Marathon, despite an attempt by an official to physically remove her from the race.

Cultivating a Holistic Approach

As we've discussed, using a whole-mind, holistic approach means shifting your mindset and expanding and contracting your field of view, often and fluidly as you work. You must be aware of which phase within your design process requires which approach, behavior, or mindset. The following three pieces of advice are intended to remind you of these in order to reinforce good design behaviors.

1. LIVE RICHLY (WITH EMPATHY)

Designing for others requires you to connect with people and understand their perspectives. Always remain interested in meeting new people, going to new places, and being open to new experiences. Cultivate a genuine curiosity about the world, especially when it comes to people. I always enjoy traveling to new countries, but you don't have to travel halfway around the world or pay for lavish events to have meaningful interactions. You can learn a lot just from your neighbors, or the people who may live just a few blocks away, but in a "different" part of town.

I remember coming out of a coffee shop late one evening during my university days and being asked for money by a man begging. Instead of brushing him off, or giving him a dollar and moving on, I sat down with him on the curb in the intersecting alley where he normally sat and just listened to his story, genuinely interested to know how he'd arrived at that time and place. It turned out that he'd been an aspiring athlete a few decades back—a football quarterback in high school, a popular man on campus—but he'd suffered an injury in his senior year. He'd never taken studying seriously—he was striving for an athletic scholarship to college, but one unlucky play ended his high-school career, and he never graduated. Without even a high-school diploma, he bounced around from odd job to odd job. Many of his high-school teammates had let him crash in their basements over the years, but slowly his relationships became strained in one way or another, and he was left on the street. There were shelters he would go to for certain meals, or on particularly inclement nights, but he enjoyed sitting on the curb and talking and bonding with people. As he grew older, he carved out a place for himself on the streets; his job was to make sure that the younger people didn't get into trouble, advising them on which shelters to go to, which alleyways were safer than others, and where they could go to count on the generosity of others.

I listened, reflecting on my own circumstances and seeing how just a few choices and a little bad luck could leave anyone houseless. And I noticed the quiet leadership he still carried with him from his days playing competitive sports. After a couple of hours, I told "Preacher Man Bill" I had to carry on and head home to study. He thanked me for just listening, and was appreciative of the respect I'd shown him. With almost a wry smile, he rummaged in his knapsack and produced an orange plastic ballpoint pen with a black cap—the kind you'd find as a free giveaway at a bank—and gave it to me, saying that he'd had such a good time "reliving the glory days," that he wanted me to have one of his most prized possessions, to "help me study." I never forgot the evening, and it did give me a new appreciation for my studies.

Experiences like this are the episodic memories that I draw on to trigger emotions relating to sacrifice and camaraderie, remixed with the compassionate concern to serve others less fortunate. Without rich memories, fueled by human emotions, your empathy and creativity won't evolve. Living richly connects you to our shared humanity, and at the same time, creates a unique palette that you bring to each design.

2. STAY CREATIVE
Cultivating a "growth mindset" is becoming an increasingly popular focus in business and management circles. In contrast to a fixed mindset, where you believe your talents are innate or god-given, a growth mindset assumes you can develop your talents through hard work, coaching, and application. Never stop learning, and never fear failure—reframe your mistakes as learning opportunities. Try again if you don't like your first attempts, and challenge yourself to keep practicing your craft so that you improve constantly. Give yourself small pet design projects to keep boosting your skills and your creative mindsets.

Design moves at the speed of life. The minute you think you've reached the summit, you've fossilized into yesterday's news. Life moves on, and with it comes new meanings out in the world. Find fellow creatives and friends whose interests you share, and whose work you admire. If you contribute equally to a common project, you'll find that your creativity becomes more fluid and responsive through collaboration. Creativity is playful, engenders trust in others, and keeps you present and responsive to the people around you. Find ways to work in that kind of environment.

3. SEE THE META (CONTEXT IS KING)
We've talked a lot about contextual awareness, and how our culture and environment shape our needs. It's only fair, then, to ask, "Why am I here?" If you're in a university setting, why are you there? What do you seek to learn, to gain? How does what you're doing contribute to the life you see five years from now? Try thinking of your life as a design project. What are the needs that drive you? Where do your aspirations fit in within your hierarchy of needs? Which of these are abstract yet actualize your potential? Which might feel good but are, in fact, superficial? There's nothing wrong with taking a little time to be spontaneous and just live in the moment, but it's also helpful to periodically identify the higher, more abstract needs that drive you.

As to career planning, if you begin with an end in mind—a purpose or mission statement—and work backward, say in five-year increments, you'll be able to identify the mentors, tools, education, and skills needed to execute each five-year plan. These plans won't be static, though; they'll shift and change based on the previous plan. I don't believe in pursuing a single passion—it might not have enough value within society to be able to sustain you. However, I do believe in cultivating many interests. You may not be passionate about a particular job or startup company, but it may touch upon several of your other interests, which will ignite your curiosity to explore further.

I hope that these three pointers provide you with some useful food for thought, and I also hope that you'll refer back to these pages frequently during your creative work. Whether you are engaging deeply with people, or looking beyond the immediate to the more implicit forces that affect your designs, I'm thankful for your commitment to learning, and optimistic about the results that you'll achieve when you make a regular practice of designing in a more holistic way.

Bibliography

Ates, Alex. "Method Acting: A Performer's Guide," Backstage, September 29, 2023, backstage.com.

Ates, Alex. "13 Types of Acting Techniques," Backstage, September 28, 2023, backstage.com.

Bartlett, John. "Chile's Atacama Desert Has Become a Fast Fashion Dumping Ground," *National Geographic*, April 10, 2023. nationalgeographic.com.

Beaty, Roger E. "The Creative Brain," *Cerebrum*, January 2020.

Berglund, Christopher. "The Neuroscience of Empathy," *Psychology Today*, October 10, 2013, psychologytoday.com.

Bhatia, Sudeep. "Inductive Reasoning in Minds and Machines," *Psychological Review*, September 2023.

BiteSize Learning. "Design Solutions to Problems with the Double Diamond Process," bitesizelearning.co.uk. Accessed August 22, 2024.

Boyatzis, Richard E., Kylie Rochford, and Scott N. Taylor. "The Role of the Positive Emotional Attractor in Vision and Shared Vision: Toward Effective Leadership, Relationships, and Engagement," *Frontiers in Psychology* 6, 2015: 670.

Bruner, Jerome. *Acts of Meaning*. Cambridge, MA: Harvard University Press, 1990.

Bustance, Jennifer. "Is Method Acting Dangerous?," Backstage, August 18, 2023, backstage.com.

College of Liberal Arts & Sciences at Illinois. "Why Is Critical Thinking Important?," las.illinois.edu. Accessed April 19, 2024.

Csere, Csaba. "2008 Tesla Roadster Road Test," *Car and Driver*, caranddriver.com. Accessed October 1, 2024.

Design Council, designcouncil.org.uk. Accessed October 14, 2024.

Ellamil, Melissa, Charles Dobson, Mark Beeman, and Kalina Christoff. "Evaluative and Generative Modes of Thought during the Creative Process," *NeuroImage* 59 (2), January 16, 2012: 1783–94.

The Enlightened Mindset. "What Part of the Brain Controls Imagination? – Exploring the Neuroscience of Creative Thinking," January 13, 2023, tffn.net.

Euro News. "French Farmers Cover Crops with Solar Panels to Battle Energy Crisis," October 2022, euronews.com.

Flam, Faye. "Inhibitions Vanish for Some with Dementia: Frontotemporal Dementia Destroys the Parts of the Brain Responsible for Judgment and Language. Studying This Disease Could Unlock Some Secrets of Alzheimer's," *Edmonton Journal*, April 1, 2002.

Fox, Taylor. "Breaking Down Different Acting Techniques: Stanislavski vs. Meisner," castingworkbook.com. Accessed October 14, 2024.

Goble, Frank G. *The Third Force: The Psychology of Abraham Maslow*. New York: Pocket Books, 1980.

Goel, Vinod, Brian Gold, Shitij Kapur, and Sylvain Houle. "The Seats of Reason? An Imaging Study of Deductive and Inductive Reasoning," *NeuroReport* 8 (5), March 1997: 1305–10.

Goleman, Daniel. "Hot to Help," *Greater Good Magazine*, March 1, 2008, greatergood.berkeley.edu.

Gross, Terry. "'Women Behind the Wheel' Explains How Cars Became a Gendered Technology," NPR, March 28, 2024, npr.org.

Guerrieri, Vince. "Baker Electric Created EV Cars in Cleveland 100 Years Ago. What Happened?," *Cleveland Magazine*, July 27, 2023. clevelandmagazine.com.

Horwitz, Simi. "What Is Sense Memory and How Should Actors Use It?," Backstage, March 19, 2009, backstage.com.

Igini, Martina. "10 Concerning Fast Fashion Waste Statistics," earth.org, August 21, 2023.

India Brand Equity Foundation. "Evolution of Micro Finance in India," ibef.com. Accessed October 29, 2024.

Interaction Design Foundation. "What Are the Gestalt Principles? — Updated 2024," interaction-design.org.

Jacobson, Danielle, and Nida Mustafa. "Social Identity Map: A Reflexivity Tool for Practicing Explicit Positionality in Critical Qualitative Research," *International Journal of Qualitative Methods* 18, January 2019.

Johnson & Wales University. "Unlocking the Power of the Mind: The Brain Region Behind Creativity and Imagination," July 6, 2023, online.jwu.edu.

Kaminska, Izabella. "Is Solar Manufacturing a Highly Automated Business?," *Financial Times*, May 18, 2021, ft.com.

Lai, Clin KY, Edith Haim, Wolfgang Aschauer, Kurt Haim, and Roger E. Beaty. "Fostering Creativity in Science Education Reshapes Semantic Memory," *Thinking Skills and Creativity* 53, September 1, 2024: 101593.

Lai, Olivia. "7 Fast Fashion Companies Responsible for Environmental Pollution in 2022," earth.org, October 15, 2022.

Lee Strasberg Theatre & Film Institute. "What Is Method Acting?," strasberg.edu. Accessed October 14, 2024.

Lerner, Ivan. "Polysilicon for Photovoltaics Growth Driven by Government Policies," Independent Commodity Intelligence Services, icis.com. Accessed October 14, 2024.

Lesley University. "The Psychology of Emotional and Cognitive Empathy," lesley.edu. Accessed September 20, 2024.

Lim, Louisa. "The Green Rush Is On In China," NPR, December 16, 2009, npr.org.

Loan, LE Thi Thanh. "The Role of Critcal Thinking in Secondary Education," *Vinh University Journal of Science* 53 (2), August 30, 2024: 165–73.

Lu, Marcus. "50 Cognitive Biases in the Modern World," Visual Capitalist, February 1, 2020, visualcapitalist.com.

Maiti, Rashmila. "The Environmental Impact of Fast Fashion Explained," earth.org, January 20, 2025.

Malara, Neha, and Noel Randewich. "Tesla Overtakes GM as Most Valuable U.S. Automaker, Short Sellers Burned," Reuters, October 24, 2019, reuters.com.

Motz, Benjamin A., Emily R. Fyfe, and Taylor P. Guba. "Learning to Call Bullsh*t via Induction: Categorization Training Improves Critical Thinking Performance," *Journal of Applied Research in Memory and Cognition* 12 (3), 2023: 310–24.

Nvisia. "What Is Agile Methodology? Benefits of Using Agile," nvisia.com. Accessed August 22, 2024.

Pillai, Unni. "Drivers of Cost Reduction in Solar Photovoltaics," *Energy Economics* 50, June 2015: 286–93.

Prado, Jérôme, Angad Chadha, and James R. Booth. "The Brain Network for Deductive Reasoning: A Quantitative Meta-Analysis of 28 Neuroimaging Studies," *Journal of Cognitive Neuroscience* 23 (11), November 2011: 3483–97.

Rietbergen-McCracken, Jennifer, Deepa Narayan. Participation and Social Assessment: *Tools and Techniques*. Washington, DC: World Bank, 1998.

Segran, Elizabeth. "Zara Built a $20B Empire on Fast Fashion. Now It Needs to Slow Down," Fast Company, July 24, 2019, fastcompany.com.

Seliger, David, "A Better World By Design: Spotlight on Panthea Lee of Reboot," Core77, October 2, 2011, core77.com.

Seyyedeh Fatemeh Seyyed Hashemi, Mehdi Tehrani-Doost, and Reza Khosrowabadi. "The Brain Networks Basis for Deductive and Inductive Reasoning: A Functional Magnetic Resonance Imaging Study," *Basic and Clinical Neuroscience* 14 (4), August 2023: 529–42.

Shellenberger, Michael. "China Helped Make Solar Power Cheap Through Subsidies, Coal and Allegedly, Forced Labor," *Forbes*, May 19, 2021. forbes.com.

Shimizu, Hirokatsu, and Ramesh Srinivasan. "Improving Classification and Reconstruction of Imagined Images from EEG Signals," *PLoS One* 17 (9), September 2022.

Singh, Satya P., Sachin Mishra, Sukrit Gupta, Parasuraman Padmanabhan, Jia Lu, Teo Kok Ann Colin, Yeo Tseng Tsai, et al. "Functional Mapping of the Brain for Brain–Computer Interfacing: A Review," *Electronics* 12 (3), 2023: 604.

Stephens, Rachel G., John C. Dunn, and Brett K. Hayes. "Are There Two Processes in Reasoning? The Dimensionality of Inductive and Deductive Inferences," *Psychological Review* 125 (2), 2018: 218–44.

Tauber, Sean, Daniel J Navarro, Amy Perfors, and Mark Steyvers. "Bayesian Models of Cognition Revisited: Setting Optimality Aside and Letting Data Drive Psychological Theory," *Psychological Review* 124 (4), 2017: 410–41.

VanderPal, Geoffrey, and Randy Brazi. "What Are Gut Feelings? Why Do They Matter?," Built In, July 9, 2024, builtin.com.

Veeravagu, Anand, and Tej Azad. "Why Brain Injury Matters In Death Row Cases," *The Daily Beast*, March 18, 2015, thedailybeast.com.

Wang, Ucilia. "The Year in Solar: A Mix of Victories and Woes," Greentech Media, December 24, 2008, greentechmedia.com.

Waseem, Ahad. "Waterfall Methodology – History, Principles, Stages & More," management.org, December 11, 2024.

Winfield, Barry. "Tested: 1997 General Motors EV1 Proves to Be the Start of Something Big," *Car and Driver*, caranddriver.com. Accessed October 1, 2024.

Index

Page numbers in italics refer to illustration captions.

A
abstract ideas 40–3
Adams, James 65
Adler, Stella 140
AEIOU (activities, environments, interactions, objects, users) 88–9, 156, 180
affinity mapping 184
Affleck, Ben 194
agile methodology 66, *66*
agrivoltaics 177
Aguilar, Francis 179
AI 119–20, *167*
air travel 99–101
Altshuller, Genrich 190
analogous research 146–7
Aston Martin DB7 6

B
Babson College 16
Baker Motor Vehicle Company 4, 5, *10*
balance 108
Bale, Christian 194
Barris, George 194
Barry, Michael 152, *182*
Batman 192–5
Begley, Ed Jr. 6
Behar, Yves 76
behavioral approaches 80–2
   contextual awareness 83, *86*, 86–7
   creativity and craft 83, 90–1
   design empathy 83, 84–5
   entrepreneurial sustainability 83, 102–5
   rapid iteration 83, *98*, 98–101
biases 132
   bandwagon effect 133
   curse of knowledge 133
   false consensus 133
   fundamental attribution error 133
   halo effect 133
   in-group favoritism 133
   moral luck 133
   self-serving bias 133
   spotlight effect 133
Boyatzis, Richard 17
brain function 17, 25
   brain control networks 29
   creativity 30–1
   deductive and inductive reasoning 28
   Gestalt theory 110
brainstorming 65, 120, 185
British Design Council 67
Brown, Brené 130
Bruner, Jerome 44–5, 148
business development 103–5
business leadership 16–23, *37*
   effective leadership 20
business viability 64, *64*, 114, 115

C
car dashboard design 95
Case Western Reserve University 16
Castiglione, Baldassare *40*
ChatGPT 120
Choi, Christina 160–7
closure 111
Clurman, Harold 140
cognitive bias 101
cognitive dissonance 45
cognitive empathy 84, 85, 130–2
colors 93
commercial design 77–9
Common to Qualifier 182–3
compassionate concern 84, 85, 139, 144–5
concrete ideas 40–3
Conran, Sir Terence 167
contextual awareness 2, 83, *86*, 86–7, 158–9
   design and business strategies 196–7
   expanding your frames of view 168–71
   organizing human needs 180–1
   timelines and mapping 186–9
   trends, technology, and culture 176–8
Continental Tire 177
continuation 110
contrast 109
COVID pandemic 22
craft 83, 90–1
Cranbrook Academy of Art, Michigan *172*, 199
Crawford, Cheryl 140
creative mindsets 38, 39
creativity 83, 90–1
critical mindsets 38, 39
Cubism 40–1
culture 176–8

D
Day-Lewis, Daniel 140, *140*, 143
De Niro, Robert 140, 142
deductive reasoning 17, 25, 27–9
Delco 9
Dench, Judi *140*
design 199, 200
   living with empathy 200–1
   seeing your own context 201
   staying creative 201
design empathy 83, 84–5, 130
design process 46–51, 60–1
   mindsets mapped to design phases 63
   phases of the design process 54, 62
   summary of the product-design process 62–3
   design-process example 52–4

design strategies 196, 197
design thinking 64, 64–5, *65*, 114
Dietz, Doug 115–16
Diffusion of Innovations 103, *104*
digital file storage 186–7
Double Diamond 67, *67*
Duchamp, Marcel *Nude Descending a Staircase (no. 2)* 40–1, *41*
Dunning–Kruger effect 101

E
E-Lab 89
Eames, Charles and Ray *172*
Eberhard, Martin 8–10
Ekman, Paul 84, 130
electric vehicles 4, 5–10
elements of art 92–6
emotional empathy 84, 85, 138–9
empathy 39, 112–13
   cognitive empathy 130–3
   dictionary definition 130
   empathy map 126–7, *127*, 180
   narratives and interviewing 148–51
   needs versus solutions 122–5
   positionality 132–3
emphasis 109
entrepreneurial sustainability 83, 102–5, 176, 197
era analysis—material culture 192–5

F
fashion industry 47–51, 54
fast (Type I) thinking 25–8, *28*
Ferrari 9, 10
figure/ground 111
Finger, Bill 192
Ford, Henry 5, 145
Ford Motor Company 6–8, 9, 10, 115
forms 92
frames of mind and frames of meaning 32–7
   combination 35
   concrete versus abstract 40–3
   deductive and inductive reasoning 33, 36–7
   frames of meaning 44–5
   frames of mind 38–9
   perspective 35
   verbal and visual approaches 34
frames of reference 168–71
Fredrickson, Barbara 17
Freeman, Morgan 194
Fresh Air *151*
Freud, Sigmund 128
frog 67
Fuseproject 76

G
games consoles 77–9
GE Healthcare 115–16

General Motor 6–8, 10
　General Motor EV1 6, 7, 9–10
Georgia Institute of Technology, Atlanta 160, 161
Gestalt theory 93, 107
　perceptual grouping 110–11
Gibson, Mel 6
Gmail 66
Goleman, Daniel 84, 130
Goodyear 177
Google 66
Gross, Terry 151

H
Hanks, Tom 6
harmony 109
Harvard Business School 179
Hasso Plattner Institute of Design 114
hierarchy & principles of design 106
　Gestalt theory 110–11
　principles of design 108–9
　visual hierarchy 107
Hoffer, Eric 24
holistic approach to design 199–201
home decor 52–3, 53, 54
Honda EV Plus 6
hospital operating rooms 146–7, 147
hues 93
human values (usability and desirability) 64, 64, 114, 115
human-centered design (HCD) 64, 65, 114
humanitarian design 76–7

I
ideas 40–3
　organizing our ideas 134–5
IDEO Product Development 6, 99, 114
　Big Iteration cycle 98
　Kelley, David 114–21
imagination 30
inductive reasoning 17, 25, 27–9
　brain activated during inductive reasoning 28
intentional states 44–5
　narratives 148–9
interviewing 148–9, 149, 151
　open-ended, semi-structured interviews 152–5
intuition 31
iteration 83, 98, 98–101

J
Jack, Tony 17
Jacobson, Danielle 136
Jobs, Steve 162
Journal of User Experience 160
journey mapping 156–7, 180–1

K
Kahn, Matt 199
Kaiser 118–19
Kane, Robert 192
Kazan, Elia 140
Keaton, Michael 193
Kelley, David 64, 114–21
　Creative Confidence 115
Kettering, Charles 5
Khan, Matt 120
Kimberley-Clark 67
Kinsella, Sophie *Confessions of a Shopaholic* 45
Koffka, Kurt 110
Köhler, Wolfgang 110
Korean Institute of Design Promotion 161

L
laptops 8, 76–7
lead-acid batteries 6–8, 9
leadership 16–23, 37
　effective leadership 20
Lego Batman Movie, The 195
Lewis, C.S. 37
Life 4
lines 92, 106
lithium-ion batteries 9–10
Lotus Elise 9–10

M
Mandela, Nelson 139
Maslow, Abraham 128
　hierarchy of needs 128–9, 180
McKim, Robert 65
Meisner, Sanford 140
method acting 139, 140–3, 1180
microfinancing 77
Microsoft 67, 77–9
mind mapping 120
　quick tips for mind mapping 135
mindsets 24–6, 38–9
　cognitive empathy 130–2
　deductive reasoning 27–9
　mindsets mapped to design phases 63
　frames of mind 38–9
　imagination, creativity and intuitive approaches 30–1
　inductive reasoning 27–9
　rational vs. intuitive approaches by discipline 25
Moggridge, Bill 8, 114
MRI machines 115–16
Musk, Elon 10

N
Narayan, Deepa *Participation and Social Assessment: Tools and Techniques* 70
narratives 148–9, 151
　contradiction 151
　normative 151
　success and failure 151
NASCAR pit crews 146–7, 147
NEA (negative emotional attractor) states 17–22
　effective leadership 20
needs 122–5
　Common to Qualifier 182–3
　Maslow's hierarchy of needs 128–9, 180
Negroponte, Nicholas 76
Nehru, Jawaharlal 170
neuroscience 17–20, 25
　imagination, creativity and intuitive approaches 30–1
　mindsets mapped to design phases 63
Nielsen, Jakob 156
　Usability Engineering 165
9 Windows method 190–1, 197
Nintendo 77–9
NN/g 156–7
Norman, Don 156

O
observation 41–3
One Laptop Per Child (OLPC) 76–7
open-ended, semi-structured interviews 152–3, 180
　advice for conducting interviews 155
　building rapport 153–4
　grand tour 154
　interview structure 152
　introduction and kickoff 153
　reflection and wrap-up 154–5
overview of design processes 64
　advice on design process 69
　agile methodology 66, 66
　case study in commercial design 77–9
　case study in humanitarian design 76–7
　design thinking 64, 64–5, 65, 114
　Double Diamond 67, 67
　human-centered design (HCD) 64, 65, 114
　Three "D" process 67
　waterfall 68, 68–9

P
Panasonic 9
Papert, Seymour 76
parents and children 44
participation strategy 74, 180
　key tips 75
　participation strategy timeline 74, 75
Patnaik, Dev 152, 182
Pattinson, Robert 194

PEA (positive emotional attractor)
  states 17–22
  effective leadership 20
perceptual grouping 110–11
PESTLE 179
physical mapping 188–9
Pirelli 177
PlayStation (Sony) 77–8
Porsche 9, 10
positionality 132, 180
  sample cognitive biases 133
  social-identity maps 136–7
Powers of 10 172–5
Powers of Ten (film) 172
prägnanz 111
principles of design 108
  balance 108
  contrast 109
  emphasis 109
  harmony 109
  proportion 108
  rhythm 108
problem solving 24, 162–4
  frames of mind and meaning 32–7
Proctor & Gamble 67
product design 12
proportion 108
proximity 110

Q
QR codes 13
  analogous research 147
  Common to Qualifier 183
  empathy maps 127
  era analysis—material culture 195
  hierarchy and the principles of design 111
  journey mapping 157
  Maslow's hierarchy of needs 128
  method acting exercises 143
  9 Windows method 190
  open-ended, semi-structured interviews 155
  participation strategy 75
  Powers of 10 175
  social-identity maps 136
  stakeholder matrix 72
  2x2 matrices 185
  weather reports 96

R
Range Rover 95
Raphael 40
rapid iteration 83, 98, 98–101
Rauch & Lang and Baker Electrics 5
Rembrandt van Rijn *Self-Portrait* 40, 40
rhythm 108
Rietbergen-McCracken, Jennifer
  Participation and Social Assessment 70

Robinson, Rick 89
Roger, Everett 103, *104*
Roth, Bernard 64
Royal College of Art (RCA), London 160, 162

S
Sacks, Jonathan 97
saturation 93
Seligman, Martin 17
shapes 92
similarity 110
Skinner, B. F. 128
slow (Type II) thinking 25–8, *28*
Snow, John 188
  Soho cholera outbreak map *188*
Snyder, Zack 194
social-identity maps 136–7
solar panels 177–8
Sony 67, 77–8
spaces 93
stakeholder matrix 70–3
Stanford University, California 65, 66, 114, 182
  Stanford d.school 64, 114, 152
Stanislavski, Konstantin 140
startups 104–5, *105*
stories 148–9, 151
Strasberg, Lee 140
surgical equipment 117–18
Swank, Hilary 142
Switzer, Kathrine 199
sympathy 130

T
Taylor, Scott 16–23
technology 176–8
  feasibility 64, *64*, 114, 115
technology-adoption curve *104*
  see Diffusion of Innovations
terrorism 195
Tesla 5, 8, 9–10
textures 93
Three "D" process 67
timelines 186–7
tools 13
  AEIOU 88–9
  affinity mapping 184
  analogous research 146–7
  Common to Qualifier 182–3
  elements of art 92–6
  empathy map 126–7, *127*
  era analysis—material culture 192–5
  hierarchy and principles of design 106–11
  journey mapping 156–7
  Maslow's hierarchy of needs 128–9, 180
  method acting for design 140–3
  mind maps and social-identity

maps 134–7
9 Windows method 190–1
open-ended, semi-structured interviews 152–5
participation strategy 74–5
PESTLE 179
Powers of 10 (10x) 172–5
stakeholder matrix 70–3
2x2 matrices 185
Toyota Corolla 10
Toyota Prius 6, 9
Toyota Production System 182
trends, technology, and culture 176–8
  economic factors 179
  environmental factors 179
  legal factors 179
  political factors 179
  social factors 179
  technological factors 179
TRIZ (theory of inventive problem solving) 190
2x2 matrices 185
  mapping stakeholders on 2x2 matrix 72–3

V
values (colors) 93
visual hierarchy 93, 94–5, 107, 108, 135
Volkswagen 176–7

W
Ward, Burt 193
Warner Brothers 193–5, *195*
washing machine design 94–5
waterfall 68, 68–9
weather reports 96
Wertheimer, Max 110
West, Adam 193
Wii (Nintendo) 77–9
Windows Vista 67
Windows XP 76

X
Xbox 360 (Microsoft) 77–9

Y
Yusuf, Hamza 151

Z
Zoom conferencing 22

Picture Credits

All illustrations by Courtney Garvin

pp.1, 2, 52, 53L, 57, 60, 113, 145, 208 Wayne K. Li
p.4 Alamy
p.5 Pictorial Press Ltd / Alamy Stock Photo
p.7TR Brain Photography / Wikipedia
p.7B © 2024 The Ford Motor Company
p.8 Tramino / Getty Images
p.16 Scott Taylor
p.28 Prof. Mehdi Tehrani-Doost, Research Center for Cognitive and Behavioral Sciences, Tehran University of Medical Sciences, Tehran
p.29 Jezper / Shutterstock
p.40 Andrew W. Mellon Collection / National Gallery of Art, Washington
p.41 Philadelphia Museum of Art / The Louise and Walter Arensberg Collection, 1950 / Bridgeman Images / © Association Marcel Duchamp / ADAGP, Paris and DACS, London 2024
p.42 Jasminko Ibrakovic / Alamy Stock Photo
p.44–45 shutterpix / Shutterstock
pp.46–47, 51 Yastrebinsky / iStockphoto
p.49 dpa picture alliance / Alamy Stock Photo
pp.50–51 Martin Bernetti / Getty Images
p.53TR nikkytok / Shutterstock
p.53M Olga Bogatyrenko / Shutterstock
p.53BR Stockphoto-graf / Shutterstock
p.55 Courtney Garvin
p.58 Igor Grochev / Shutterstock
p.69 el_cigarrito / Shutterstock
p.76 Mike McGregor / Wikipedia
p.79 ST-images / Alamy Stock Photo
pp.85, 138L Pheelings media / Shutterstock
pp.85, 138R Vitalii Matokha / Shutterstock
p.86 Pixel-shot / Ken Hurst, Alamy Stock Photo / Trong Nguyen, Shutterstock / Shutterstock
p.89T mavo / Shutterstock
p.89TM Wavebreak Media Premium / Alamy Stock Photo
p.89M Monkey Business Images / Shutterstock
p.89BM Thongsuk Atiwannakul / iStockphoto
p.89B PeopleImages.com - Yuri A / Shutterstock
pp.90–91 Shutterstock: neuralsuperstudio, Stock Holm, EngravingFactory, Tatevosian Yana, BongrakArt, Light Stock, idea _Photo, AspctStyle, Jovica Varga, Photoongraphy, Photoongraphy, BongrakArt, so.ni.ka, MaraZe, Natalie_3dArt

p.94 NosUA / iStock
p.95 © Jaguar Land Rover Automotive PLC
p.96 Eric Mischke, SpiffyJ, ronemmons / iStockphoto
p.99 Feel good studio / Shutterstock
p.102 Magnifical Productions / iStockphoto
p.114 David Kelley
p.123 STEKLO / Shutterstock
p.139 My Ocean Production / Shutterstock
p.140 PA Images / Alamy Stock Photo
p.143T Miramax/Dimension Films/Kobal / Shutterstock
p.143M Moviestore / Shutterstock
p.143B Phantom Thread / Alamy
p.146 Icon Sportswire / Getty Images
p.147 Morsa Images / Getty Images
p.149 Thierry Falise / Contributor
p.150 Bill Cramer / Wonderful Machine
p.160 Christina Choi
p.166T SolStock / iStock
p.166M Koji_Ishii / iStcok
p.166B FG Trade / iStock
p.172 © Eames Office, LLC. All rights reserved
p.174 Shutterstock: Ground Picture, Grusho Anna, Photo Smoothies, Roman Kosolapov, alice-photo, William Perugini, NicoElNino; iStock: Sanya Kushak
p.177B Bloomberg / Getty Images
p.178 Jenson / Shutterstock
p.191TL The Print Collector / Alamy Stock Photo
p.191TM Joseph Hendrickson / Shutterstock
p.191TR Tatyana Ledneva / Shutterstock
p. 191BL Panther Media GmbH / Alamy Stock Photo
p.191BM AlexLMX / Shutterstock
p.191BR Courtney Garvin
pp.192–93 TOP ROW L–R Fox/Abc/Kobal/Shutterstock, Entertainment Pictures / Alamy Stock Photo, Collection Christophel / Alamy Stock Photo, Warner Bros/Everett/Shutterstock
BOTTOM ROW L–R ABC/Shutterstock, Snap/Shutterstock, TCD/Prod.DB / Alamy Stock Photo, Moviestore Collection Ltd / Alamy Stock Photo
p.195 Warner Bros/Everett/Shutterstock
p.198 Walter Iooss Jr. / Getty Images
p.199 LANDMARK MEDIA / Alamy Stock Photo

Acknowledgments

Dedicated to my parents, Hsueh Ming, Shu Nu, and Hsiu Chun, who taught me a love for science, art, and service. And to my family, Jubi, Asher, and Sophia, and my siblings Al and Sandy and their families.

I'm grateful for the educational experiences that shaped me; design programs at the University of Texas at Austin and Stanford University, which grounded and inspired a holistic approach to design; and the College of Creative Studies in Detroit, for inspiring my attention to detail.

I'm also grateful for the following professional design work: Corporate Design Studios at Ford Motor, Electronics Research and Interaction Design at Volkswagen, home decor and interior design at Williams Sonoma, and product development at IDEO Product Development and Design Edge.

Thanks go to Jim and Kendall Oliver, for their generosity and my chair at the Georgia Institute of Technology's Industrial Design and Mechanical Engineering programs, without which none of my research into design would have been possible. I also benefited from the opportunity to teach product design and give back through Stanford's Product Design program, which ignited my passion and revealed my potential in academia.

And finally, I acknowledge the students, friends, and colleagues who make this field of design my life's passion and privilege. My thanks to you, for helping me become the design educator I am today.

RIGHT: A collage of creative output—scribbles over a year from Asher (age 5) and Sophia (age 3).